JOHN MURRAY SCIENCE PRACTICE

teaching secondary
PHYSICS

JOHN MURRAY SCIENCE PRACTICE

teaching secondary
PHYSICS

Editor: David Sang

JOHN MURRAY

Titles in this series:

Teaching Secondary Biology   0 7195 7637 7
Teaching Secondary Chemistry   0 7195 7638 5
Teaching Secondary Physics   0 7195 7636 9

Photographs are reproduced courtesy of: **cover** David Parker for ESA/CNES/Arianespace//SPL; pp. **47**, **70** & **196** Andrew Lambert.

The publishers have made every effort to contact copyright holders. If any have been overlooked they will make necessary arrangements at the earliest opportunity.

First published in 2000
by John Murray (Publishers) Ltd, a member of the Hodder Headline Group
338 Euston Road
London NW1 3BH

Reprinted 2000, 2001, 2002 (twice), 2003 (twice), 2004, 2005

Illustrations by Art Construction
Layouts by Amanda Easter

Typeset in 12/13pt Galliard by Wearset, Boldon, Tyne & Wear
Printed and bound in Malta

A catalogue entry for this title is available from the British Library

ISBN 0 7195 7636 9

Contents

Contributors

Martin Hollins is Principal Subject Officer for science at the Qualifications and Curriculum Authority. He has taught physics in secondary schools in London, Surrey and Malawi. He has been involved in in-service and initial teacher education for many years and has directed several major curriculum developments for use in both secondary and primary schools. He has written widely in science education from the primary phase to advanced level physics.

Bob Kibble taught in London for 22 years and now lectures in physics and science education at the University of Edinburgh. He was a Chief Examiner for GCSE Physics and an inspector of Further Education colleges. His professional interests include writing curriculum materials, promoting astronomy education and delivering workshops for teachers. He is Deputy Editor of *Physics Education*.

Robin Millar taught physics and general science in comprehensive schools in the Edinburgh area before moving to the University of York where he is now Professor of Science Education. He teaches on the Science PGCE course and various higher degree programmes. He has been involved in several major curriculum projects, including Salters' Science and Salters' Horners' A-level Physics. His research interests include students' understanding of science ideas and the role of practical work in science teaching.

Jonathan Osborne is a senior lecturer in science education at King's College, London, where he has been since 1985. Prior to that he was an advisory teacher with the ILEA and taught physics and astronomy in secondary schools for nine years. He has an MSc in Astrophysics and has regularly run in-service courses on teaching the topic of the Earth and Beyond. He has also undertaken extensive research for the SPACE project on young children's understanding of this topic and written articles for the *School Science Review* and *Primary Science*.

David Sang worked as a research physicist for six years before becoming a teacher. He has taught physics for 13 years, in a comprehensive school, a sixth-form college and a polytechnic. He now works on a variety of curriculum development projects, as well as writing and editing textbooks and other teaching materials for pupils aged 11 to 18.

Joan Solomon is now professor in the Centre for Science Education at the Open University. Previously she was lecturer in research at Oxford University and taught physics to advanced level for more than 25 years in various schools in London and elsewhere. She also initiated projects on the teaching of STS (Science, Technology and Society), for which she was Chief Examiner for about 12 years. She has written several books on the teaching of science and many articles on science education research.

Acknowledgements

This book has benefited from comments and contributions from many practising teachers of physics. In particular, the authors and editor are very grateful to the following for their advice during its preparation:

Graham Booth
Chris Butlin
Nigel Heslop
Sarah Howes
Averil Macdonald
Alan Pickwick
Liz Swinbank
Nigel Wallis
John Warren

The editor is also grateful to the following for their support and encouragement in this project:

Jane Hanrott
Andrew Hunt
David Moore
Mick Nott
Mary Whitehouse
Catherine Wilson

The Association for Science Education acknowledges the generous financial support of ESSO and the Institution of Electrical Engineers (IEE) in this project. Both ESSO and IEE provide a range of resources for science teaching at primary and secondary levels. Full details of these resources are available from:

ESSO Information Service
PO Box 94
Aldershot
GU12 4GJ
(tel: 01252 669663)

IEE Educational Activities
Michael Faraday House
Six Hills Way
Stevenage
SG1 2AY
(e-mail: nsaunders@iee.org.uk)
(web site: www.iee.org.uk/Schools)

Introduction

David Sang

This book is one of a series of three ASE handbooks, the others being parallel volumes on teaching biology and chemistry. It has been written by a team of experienced physics teachers, assisted by comments and contributions from many others. Our aim has been to outline a pragmatic approach to the teaching of physics, against a background of discussion of some of the issues involved. We have at all times kept in mind the needs of a teacher confronted with the task of teaching a specific topic in the near future, and tried to address the question, what does such a teacher need to produce a series of effective lessons?

Who is the book for?

In writing this book, the authors have kept in mind a range of likely readers:

- biologists, chemists and science generalists who find themselves teaching parts of the physics curriculum;
- new or less experienced physics teachers, although almost every physics teacher should find much of value;
- student teachers, their tutors and mentors;
- heads of department who need a resource to which to direct their colleagues.

While the current UK syllabus requirements have been taken into account, the content does not closely relate to any one curriculum. We expect that this book will be appropriate to secondary physics teachers throughout the UK, and elsewhere.

What should you find in the book?

We expect you, the reader, to find:

- good, sensible, reliable and stimulating ideas for teaching physics to 11–16 year olds;
- suggestions for extending the range of approaches and strategies that you can use in your teaching;
- things to which pupils respond well; things that fascinate them;
- confirmation that a lot of what you are already doing is fine – there often isn't a single correct way of doing things.

How can you find what you want?

We have divided the physics curriculum into six main areas. From the titles of the chapters as listed on the Contents page, you will see roughly what each chapter deals with. A more detailed indication is given by the 'content map' at the beginning of each chapter. This divides each chapter into about half a dozen sections. For specific topics, consult the index which contains the likely terms you might want to look up.

What is in each chapter?

Each chapter contains:

- a content map which divides the chapter up into shorter sections;
- an indication of possible teaching routes through the content of the chapter;
- a brief section on what pupils may have learned or experienced about the topic earlier in their science education;
- an outline teaching sequence showing how concepts can be developed throughout the 11–16 phase;
- warnings about pitfalls;
- information about likely pupil misconceptions;
- helpful advice about practical work and apparatus, e.g. how to prevent things from going wrong;
- more specific equipment notes at the end of some chapters;

- issues to do with safety (highlighted by the use of hazard icons in the margin);

- suggestions for the use of ICT (highlighted by the use of floppy disk icons in the margin);
- suggestions about the use of books, videos and other resources;
- opportunities for the teaching of the applied or ethical aspects of physics;
- links with other areas of physics or science.

Each chapter is, however, different. These differences of emphasis are a reflection of the different topics covered. For example, the teaching of energy has been hotly debated over the years, and several different approaches are enshrined in syllabuses. Chapter 1 therefore aims to help teachers to understand the issues underlying these approaches. By contrast, specialists and non-specialists alike may feel nervous when required to teach about radioactivity. Chapter 6 aims to show how you can tackle this topic with confidence through an understanding of the physics involved.

Safety first!

As a teacher working in a science laboratory, you have responsibility for the safety and well being of your pupils. There are relatively few serious hazards in the teaching of physics at secondary level; particular hazards are highlighted throughout this book.

Organisations such as CLEAPSS and ASE have worked hard to identify possible hazards and to disseminate information about how to minimise dangers. However, the fact that there are relatively few serious accidents in school science laboratories reflects the care with which teachers prepare for practical lessons by considering possible hazards and minimising risks.

Before planning any practical activity you should:

- identify possible hazards in materials, equipment and procedures;
- try out the activity beforehand yourself so that it is not happening for the first time with a class;
- produce a risk assessment, which may well be included in your department's scheme of work;
- stick to what you have planned to do – don't be tempted to try new ideas on the spur of the moment.

You should be aware of the safety recommendations from CLEAPSS and ASE, and you should also check if there are any limitations on practical work made by your Local Education Authority. If you are using activities from a published scheme of work, it is likely that the teacher's guide will contain risk assessments; you will need to decide whether these are adequate for your own situation, or whether they need adapting.

Specific points about safety in school physics laboratories

Teaching pupils about safe working in the laboratory should be the subject of a policy that extends across the science department. You should be aware of the procedures that pupils are being taught in their chemistry and biology lessons. In particular, you should know whether standard risk assessment forms are in use. You will find examples of these in the companion volume to this book, *Teaching Secondary Chemistry*. You could ask pupils to complete these forms for one or two selected physics activities, to show that the same approach to safety applies here, too.

You should also make yourself aware of the safety features in each laboratory that you work in. Where is the trip for the mains electricity supply? Where can you turn off the gas or water supplies? What fire safety equipment is provided? What escape route should pupils use in the event of a fire?

In the event of an accident it is important for you to know the correct procedures for first aid and emergency treatment. Within the staff of the school there should be at least one person trained in first aid and provided with a fully-stocked first aid box. Very serious accidents involving loss of limbs, damage to eyes, and unconsciousness have to be reported to the Health and Safety Executive. There should also be a system for recording minor accidents, such as small burns and cuts, in case reference needs to be made later to what happened and what treatment was given.

Accidents can occur through teachers and pupils failing to wear eye protection or to use safety screens. Both may impede pupils' view of an activity, particularly a demonstration; however, don't be tempted to do without them. Ensure that pupils keep eye protection on, and that they sit where they are viewing the equipment through the screen, not around the side. Consider carrying out a demonstration several times with smaller groups, rather than just once with the whole class. You will need to provide something worthwhile for the others to do meanwhile.

Special regulations apply to the use of lasers and radioactive materials. Make sure you are familiar with these before you tackle these parts of the curriculum.

Pupils may be nervous about the use of some equipment, such as power supplies. While these are designed to be safe, you should respect their caution. At the same time you should help them to learn how to use such equipment safely.

Useful contacts

CLEAPSS School Science Service, Brunel University, Uxbridge, Middlesex UB8 3PH (tel: 01895 251496; fax: 01895 841372; e-mail: science@cleapss.org.uk).
(Most local authorities subscribe to this service and schools are able to seek advice on safety matters for no charge.)

SSERC, St Mary's Land, 23 Holyrood Road, Edinburgh EH8 8AE (tel: 0131 558 8180; fax: 0131 558 8191; e-mail: sserc@mhie.ac.uk; web site: www.vtc.scet.com/links/sserc).

Association for Science Education, College Lane, Hatfield, Herts AL10 9AA (tel: 01707 283000; fax: 01707 266532; e-mail: ase@asehq.telme.com).
(Notes on safety issues are frequently included in the ASE members' publication *Education in Science*.)

A consultant to CLEAPSS has read this text and confirms that, in the draft checked, the identification of hazards and the precautions given either conform with published general risk assessments or, if these are not available, are judged to be satisfactory.

1 *Energy*

Robin Millar

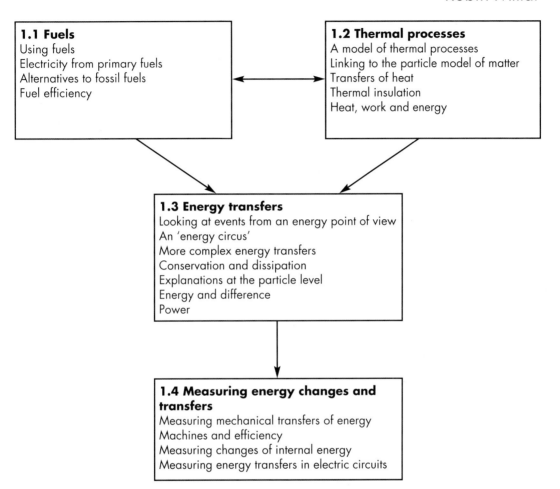

1.1 Fuels
Using fuels
Electricity from primary fuels
Alternatives to fossil fuels
Fuel efficiency

1.2 Thermal processes
A model of thermal processes
Linking to the particle model of matter
Transfers of heat
Thermal insulation
Heat, work and energy

1.3 Energy transfers
Looking at events from an energy point of view
An 'energy circus'
More complex energy transfers
Conservation and dissipation
Explanations at the particle level
Energy and difference
Power

1.4 Measuring energy changes and transfers
Measuring mechanical transfers of energy
Machines and efficiency
Measuring changes of internal energy
Measuring energy transfers in electric circuits

♦ *Choosing a route*

The idea of energy is not at all obvious. Nor is the idea of its conservation. Indeed our intuition is the opposite: that something is lost, or used up, when changes occur.

The scientific idea of energy took shape only during the 1850s. Since then, however, the word 'energy' has entered everyday language. Most pupils by the age of 11 have heard of energy and are able to use the word appropriately in ordinary conversation. 'Energy', however, has a much less precise meaning in everyday conversation than in science. Indeed the everyday usage can even seem to contradict the scientific one. For example, in everyday contexts we may talk about 'using up' energy, or about 'energy consumption', whereas science says that energy is always conserved. These everyday ideas can make it more difficult to teach the scientific idea of energy.

There has been more debate and discussion, in journals like *School Science Review* and *Physics Education*, about the teaching of energy than any other physics topic. Much of this has centred on the language we should use when talking about energy. People have argued for and against the use of 'forms of energy'. If you choose to avoid 'forms', then you have to talk about energy being 'transferred', rather than 'transformed' or 'converted'. Others have argued that we should not use 'heat' as a noun, but talk about 'heating' as a process. It would be nice, in this chapter, to be able to outline the consensus view on these issues, but there simply isn't one. Indeed the 'official' curriculum documents in England and Wales and in Scotland take two quite different approaches. The teaching sequence suggested in this chapter uses ideas from both of these. Where issues of this sort arise and choices have to be made, these are discussed in a series of boxes within the main text.

Ideas about energy are closely inter-related, so it is important to think carefully about the order in which you introduce them in your teaching. We begin by looking at fuels and how we use them. This builds on pupils' everyday knowledge and links directly to everyday contexts. Most of Section 1.1 would be suitable for a lower secondary course, though the last sub-section (*Fuel efficiency*) might be better left until upper secondary stage, after Sections 1.2 and 1.3 have been covered.

Section 1.2 looks at how we can develop pupils' understanding of thermal processes. These ideas are then used in Section 1.3, which looks at the idea of energy more generally. Most of this material would normally be introduced at lower secondary stage.

Section 1.4 then discusses a more quantitative treatment of energy ideas. This is suitable for upper secondary classes, especially for pupils who may wish to go on to more advanced courses in science. With less academic pupils at upper secondary stage, it may be more profitable to revisit some of the ideas about fuels towards the end of Section 1.1, looking in more depth at their implications for society and for personal lifestyle choices.

1.1 Fuels

When people talk (or write in newspapers and magazines) about 'energy consumption', or about 'using' or 'using up' energy, they are often using the word 'energy' to mean 'fuel'. We *should* 'save fuels' where possible. Fuels *are* in limited supply and need to be used with care. When we use a fuel to make a process happen, the fuel *is* used up. So we can extend and develop pupils' ideas about fuels, by building on their prior ideas and intuitions. According to the Oxford English Dictionary, a fuel is 'material for burning, combustible matter for fires'. But rather than try to define 'fuel' precisely, it is better simply to give some examples. Starting from the obvious ones, like coal, wood, oil and gas, we can extend the dictionary definition to include other kinds of fuel, and to incorporate electricity within the general notion of a fuel.

Previous knowledge and experience

Although they may not have had any formal teaching about fuels or energy at primary school, pupils entering secondary school will have many ideas about fuels and how we use them. Most will be familiar with the use of fuels like gas, oil or coal to make (and keep) places warm, and for cooking. They will also know that cars, lorries and other forms of transport need a fuel of some kind to make them 'go', and that petrol and diesel are two fuels that are used. By age 11, pupils will also know that electricity can be used for heating and cooking, and for some kinds of transport (e.g. trains), and they may think of it as simply another kind of fuel. They will also be familiar with devices that need batteries to make them work, such as torches, radios and personal stereos, and will know that these batteries have a limited lifetime.

A teaching sequence

Using fuels

A topic on fuels at lower secondary stage can help pupils to organise and extend their informal knowledge. A good way to begin is by eliciting pupils' prior knowledge, for example, by collecting information on what is needed to make a range of things 'work' or 'go', such as cars, trains, aeroplanes, cookers, the heaters in the classroom or at home, portable radios, and so on. In this way, you can build up a list of fuels from the pupils' own prior knowledge. Their suggestions are likely to include electricity at this stage.

You might then follow this up by looking at some data on the amounts of each of the major fuels used in the UK for different purposes (domestic, transport, industry and commerce, etc.) or in different historical periods; or on fuel use in different countries. Activities like these provide good opportunities for pupils to practise and develop their ICT skills, using spreadsheets, databases and graphical presentation packages.

You might then introduce the idea of *fossil fuels*, explaining how these are thought to have formed (so that stocks are therefore fixed and finite), and looking at the rate at which they are being used up. This could involve looking at patterns of consumption of fossil fuels over time, amounts used in different countries, projections of future use and lifetimes of stocks. Pupils might produce an information leaflet (for a particular audience), a tape–slide sequence or a poster, explaining the formation of fossil fuels and their value to us.

Electricity from primary fuels

It is quite likely that pupils will suggest electricity as an example of a fuel. You might show pupils that batteries contain chemicals, and explain that these are used up when the battery is in use. In advance of the lesson, cut a 1.5 V dry cell in half using a hacksaw. (Do not use an 'alkaline' cell or a nickel–cadmium rechargeable.) Clamp the cell firmly in a vice and cut it vertically to produce two half cylinders.

The cut edges of the metal case may be sharp, so make sure pupils take care when handling it.

Although there is nothing dramatic to see, you can point out that the cell contains at least two different chemicals. Show pupils that there is nothing 'magic' about these chemicals, by demonstrating that ordinary materials can make a cell: a potato with two different metal rods inserted into it will produce an electric current (Figure 1.1). The fuel here is the chemicals in the potato and the metal rods, which are gradually used up in the process.

Figure 1.1
A potato and rods of two different metals can form a cell. Kits are available to make such a cell power a digital clock.

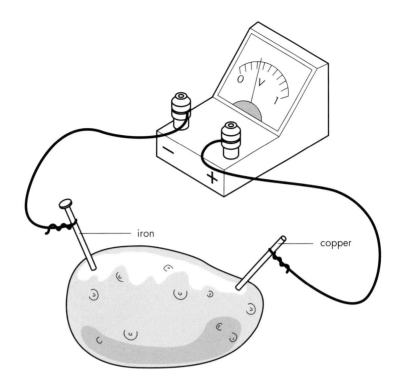

As regards to mains electricity, you should explain that this is not a 'primary fuel' but has to be generated using a primary fuel. Without going into detail about the processes involved in power stations, you should explain that most burn a fossil fuel (coal and gas being the commonest ones in the UK at present) to generate electricity which is then distributed by a network of cables and wires. So we can think of electricity as a convenient way of using a fuel 'at a distance'. Some pupils may also know that some power stations use nuclear fuel (enriched uranium or plutonium) and some are hydroelectric, using water in a high reservoir as the 'fuel'. These can be added to the list of primary fuels.

You may like to demonstrate the generation of electricity from a primary fuel, using a model steam engine to turn a dynamo and light a small lamp (Figure 1.2, overleaf).

! *Note carefully the safety guidance offered to subscribers by CLEAPSS (L214a, b, c). In particular, only a solid fuel (such as metafuel) should be used.*

By using (or imagining) long leads, the lamp can be far from the generator. So electricity allows us to do useful jobs a long way from where the primary fuel is being used up.

Figure 1.2
*Demonstrating
how a primary
fuel may be used
to generate
electricity.*

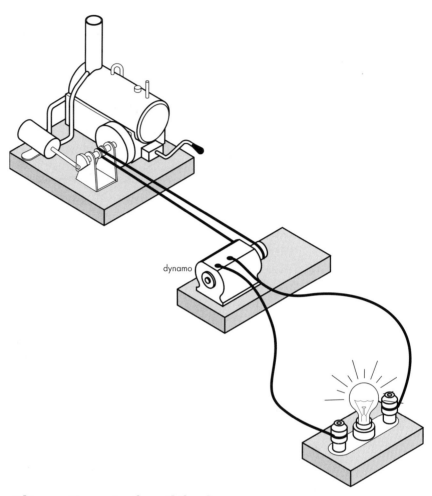

dynamo

Alternatives to fossil fuels

Many pupils, by lower secondary school, will have heard of the
idea of saving fossil fuel reserves by using alternative, 'renewable
energy sources', such as hydroelectric dams, sunlight, wind, waves
and tides. As they will not, by this stage, have learned about the
physics involved in many of these processes, the discussion should
be in outline terms only. However, we might expect pupils, at
lower secondary stage, to be able to say what the following devices
do and (in very general terms) how they work: hydroelectric
power stations (including pumped storage), passive solar heating
panels, solar cells, wind turbines, wave generators, tidal barrages.
Other renewable sources which should be mentioned are: biomass
fuels (including wood), alcohol from fermented sugar cane to
replace petrol, and methane from domestic waste. These are
essentially sustainable ways of producing a primary fuel which can
be substituted for a fossil fuel.

Pupils should learn that, although renewable sources can help us reduce our fossil fuel consumption, they also have some disadvantages. They are not always available when we want them most, they can be expensive, and some also have a significant environmental impact.

Possible activities on renewable energy sources include:

♦ Library-based research. Divide the class into groups, each finding out about a different renewable energy source, and how it may be harnessed. Each group presents its findings to the whole class.

♦ An investigation of the current obtained from a small photovoltaic cell (a model solar panel), or of the speed of rotation of a small electric motor connected to it, at different levels of illumination.

♦ Constructing working models of a wind generator or a solar heating panel. Simple designs can be found in the *Teachers' Guides* on *Wind Power* and *Solar Heating* (from the Centre for Alternative Technology) or in the *Energy Resources* module of the *Pathways Through Science* materials (Hunt & Milner, 1992). This sort of activity can, however, be very time-consuming and is perhaps best undertaken in an after-school science club.

Fuel efficiency

Another way to reduce fossil fuel use is to improve efficiency. You will need to explain that 'fuel efficiency' is a measure of how good is the 'value' we are getting from the fuel we use in any particular situation. As many of these involve heating, it may be wise to leave the idea of efficiency until pupils have met the ideas in Section 1.2.

You might then ask the class to brainstorm possible ways of using fuel more efficiently in some everyday situation. Their situations might include: designing cars (or engines) which do more miles to the litre; using buses or trains rather than private cars (more passenger-miles per litre); installing good house insulation; having showers rather than baths; using compact fluorescent light bulbs rather than filament bulbs; keeping the lid on a saucepan while boiling water. The common feature is getting the same outcome with less fuel.

Possible practical activities include:

♦ Comparing the time required to boil water (or raise its temperature by a given amount) in a beaker with and without a lid, using either a Bunsen burner (on a constant setting) or an electric immersion heater.
♦ Designing an arrangement to maximise the temperature rise of 100 ml of water in a beaker, or tin can, using a single barbecue lighter (or similar small solid fuel block). Note that good ventilation is required.
♦ Practical investigations of various domestic fuel-saving measures, using a 'model house', such as the *TASTE* model available from ASE Booksales.

♦ *Further activities*

At lower secondary stage, the depth of treatment of ideas about fuels is necessarily limited by the pupils' knowledge and experience in other areas of physics. As these ideas relate directly to choices and decisions we all have to make in our everyday lives, it is important to return to them at upper secondary stage, when the pupils have learned about electricity generation, thermal processes, radioactivity and so on.

Possible activities on fuels at upper secondary stage include:

♦ More detailed discussion of how the different types of power station work, their efficiency as energy transformers, and the use of combined heat and power (CHP) stations to improve overall efficiency.
♦ More quantitative work on trends and patterns in fuel use in the UK and other countries. This could involve library-based research and the use of public information leaflets and official statistics as information sources.
♦ Using data to explore in more depth the advantages and disadvantages of the various renewable energy sources.
♦ Debates, role plays and decision-making exercises about methods of generating electricity for a community, and about the links between fuel use and lifestyles.

1.2 Thermal processes

At lower secondary stage, we want to help pupils to think and talk more precisely about thermal processes. In order to do this, they need to have a 'model' of thermal processes in mind. Two key ideas in this model are *temperature* and *heat*. Pupils have to learn to distinguish these two ideas.

Heat or energy?

There has been much discussion and debate in the science education journals over the years about whether we should use the word 'heat' as a noun. The essential problem is that people tend to use 'heat' to mean two things: the energy contained in a hot object, and the energy transferred between two objects because of the temperature difference between them. This is a problem because these two things are not the same. (This is discussed in more detail in the sub-section *Heat, work and energy*, page 19.) Some people argue that we should therefore avoid using the word 'heat' as a noun and simply call the thing that is stored in hot objects 'energy'. The *process* of transfer of energy due to a temperature difference would then be called 'heating'.

But this also leads to some difficulties. We cannot, for example, talk about teaching pupils the difference between temperature and heat. We would have to talk instead about teaching them 'the difference between temperature and energy', which does not make much sense. More practically, it is difficult to avoid talking about 'heat' in class. Pupils are likely to use the word when discussing thermal processes even if the teacher manages to avoid it. And if we choose to refer instead to 'energy', how do we explain to the pupils what it means? Remember that they have not yet had a general introduction to energy ideas, and so their understanding of the word 'energy' is based on its everyday meaning (which is rather different from the scientific one).

Does this mean that it would be better to teach a general unit on energy before looking at thermal processes? This also

---

runs into problems because a unit of this sort would need to bring in the idea of energy being transferred to objects, making them hotter – which assumes an understanding of the difference between energy (or heat) and temperature.

As with many physics topics, there is no ideal sequence that allows us to build up the ideas logically 'from the bottom up'. Each of them depends on the other. We have to start somewhere, by introducing an idea and using it, but recognising that we will need to return to it later to explain and define it more precisely. So here we opt to use the word 'heat' at first, and choose situations where all the important energy transfers are thermal ones. Learning to distinguish clearly between the 'intensive' quantity (*temperature*) and the 'extensive' quantity (*heat*) represents a major step for many pupils. Once they have grasped this, we can go on to consider situations that involve energy transfers by work, and make the distinction between heat and (internal) energy. If, however, you prefer to avoid the word 'heat' from the outset, then anywhere it is used as a noun in Section 1.2, it can be replaced by 'energy'.

◆ *Previous knowledge and experience*

Pupils have a store of everyday experiences of things heating up and cooling down, and have ideas about these processes before they have been taught anything about them formally. Most will know that hot things left to themselves tend to cool down, whilst cold things warm up. And they also know that if you put something on or near a hot object, it gets hotter too (for example, a saucepan on a cooker hob or your hands on a warm radiator).

At primary school they may have used a thermometer to measure the temperature of a cup of hot water over a period of time, noting that its temperature drops until it is the same as the room. And they may have investigated how the temperature of iced water rises. Some may also have seen how insulation can slow this process down. Many, however, will not be able to explain these observations using scientific ideas.

◆ *A teaching sequence*

A model of thermal processes

Even though pupils may have done something similar at primary level, a useful starting point is a practical investigation of the change in temperature of a beaker of hot water and a beaker of iced water, over a period of time (Figure 1.3). One aim is to draw their attention to points that they may not have previously observed:

- that the final temperature which both hot and cold water approach is that of the surroundings;
- that the rate at which the temperature changes gets smaller as the object's temperature approaches that of the surroundings.

If you use a graph of temperature against time to present these results, remember that pupils at lower secondary stage may not have had much practice in drawing graphs of this sort. This may be an opportunity to teach some specific points about line graphs in science, such as linking points by a smooth curve, rather than joining the dots.

Figure 1.3
Looking at how hot water cools down, and cold water warms up.

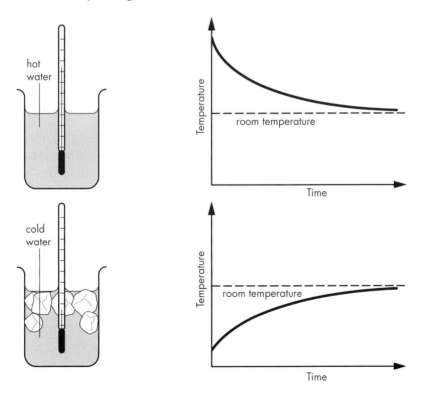

-1

Figure 1.4
A model of thermal processes.

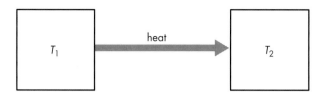

If $T_1 > T_2$, heat flows in the direction shown.

For any given pair of objects, the bigger $(T_1 - T_2)$, the greater the rate of heat flow.

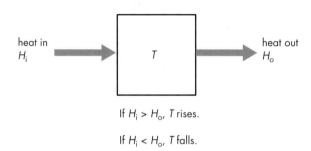

If $H_i > H_o$, T rises.

If $H_i < H_o$, T falls.

A second aim is to introduce a model which can help us to *explain* these observations. We say that heat flows *spontaneously* from regions of higher temperature to regions of lower temperature. The bigger the temperature difference, the faster the transfer of heat (if everything else stays the same). When an object loses heat, its temperature falls; when it gains heat, its temperature rises. This model is summarised in Figure 1.4, and is referred to several times in the pages which follow.

The role of the environment
To explain our observations in the activity above, we have to include the environment as part of the system. It is not always obvious to pupils that the environment plays a role, so it is worth highlighting this explicitly. A useful practical exercise is to place beakers of hot water in cold water baths of different volumes, and monitor the water temperature in both containers over a period of 20–30 minutes (Figure 1.5). In all cases, the temperature drop of the hot water in the inner beaker will be obvious. The temperature rise of the cold water bath will be quite clear when its volume is small, but gets less as the bath increases in volume. With the largest one, there may be no measurable temperature rise. Pupils can, however, accept that its temperature has risen, even if the rise is undetectable. They may then be more willing to accept that the surroundings can be a sink or a source of heat, even though their temperature may not change by a measurable amount.

Figure 1.5
Modelling the environment as a sink for heat.

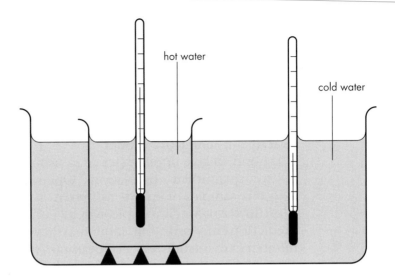

hot water

cold water

Thermal equilibrium

Another implication of the model of Figure 1.4 is that objects in thermal contact eventually reach the same temperature. This may conflict with pupils' intuitions that some materials in a room are colder than others – tiles feel colder to the feet than carpet; metal objects feel cooler than ones made of wood or plastic.

A useful activity is to get pupils to measure the temperature of objects of different materials within the laboratory. You could use blocks of wood, plastic, stone or ceramic and metals, pre-drilled with holes to accept a thermometer bulb or temperature probe, or a liquid crystal thermometer can be stuck on to the surface. Ask pupils to predict the temperature before making measurements, as this may help you understand how they are reasoning. You can explain that materials that feel cold are those which conduct heat quickly away from your hand. If you have not yet introduced the idea of conduction, you may want to keep this explanation for later. The important idea for pupils to grasp at this point is that objects which have been in the same environment for some time are all at the same temperature.

Using analogies

Although these ideas may seem simple ones, many pupils need time to sort out the difference between heat and temperature. They may find it helpful to think of temperature as a measure of the 'concentration' of something (energy), and of heat as a measure of the total amount of it in the object. Another analogy which some pupils find useful is the depth of water in a container with an inflow and an outflow. If the rate of inflow is greater than the rate of outflow, the depth will increase, and vice versa. So the water flowing in and out is like the heat entering or leaving an object, whilst the depth of water represents the temperature.

Practice in using the model

It is good to give pupils practice in using the model of Figure 1.4 to explain their observations. One simple practical activity is to mix samples of liquid at different initial temperatures. Most lower secondary pupils will be confident in stating their predictions in words, but some may have difficulties when numbers are introduced. For example, they can predict correctly that mixing equal volumes of hot and cold water will result in luke-warm water, but then predict that adding samples of water at 20°C and 60°C will result in water at 80°C. Again this can be very effective if presented as a Predict–Observe–Explain (POE) task. Ask the pupils to say what they expect, then to do it, and finally to explain their observations. You can go on to mix two parts hot with one part cold, and vice versa.

It can also be useful to discuss some 'thought experiments' with the whole class. Ask them to imagine taking a cup of water out of a swimming pool. What effect would this have on the temperature of the swimming pool and on the amount of heat it contains? How does the temperature of the water in the cup compare with that in the swimming pool? How much heat is in the water in the cup as compared with the water in the swimming pool?

One other practical task which can help pull together many of these ideas is to ask pupils to measure the temperature over 30 minutes of a sample of water in a tin can heated by a 'night-light' candle (Figure 1.6). With 50 ml of water in the can and a distance of a few centimetres between the flame and the tin, the temperature of the water will rise to around 50–60°C. Ask the pupils to explain why the temperature of the water rises initially, but later reaches a steady value. This brings in most of the key ideas of the model in Figure 1.4: the spontaneous transfer of heat from regions of high to regions of low temperature, the rise in temperature of objects to which heat is transferred, the role of the surroundings as part of the system, and steady temperature when heat is entering and leaving an object at the same rate.

Figure 1.6
Investigating how the temperature of a sample of water changes as it is heated.

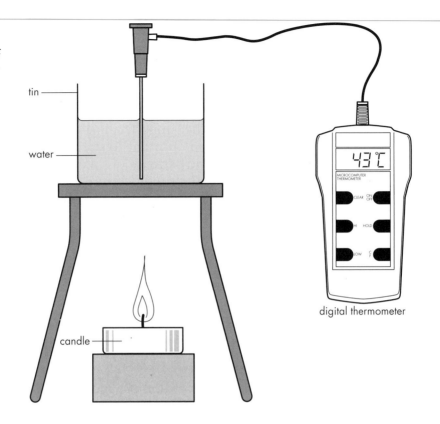

digital thermometer

Linking to the particle model of matter

You can discuss heat and temperature without using ideas about the particulate nature of matter. However, pupils are likely to have met particle ideas by lower secondary stage, and it is useful to make links between the two areas. (See *Brownian motion and the kinetic model* in Chapter 3, page 113.) The key idea here is that temperature is a consequence of the motion of the particles of a material. If we raise the temperature, we make the particles move faster; and, conversely, if something makes the particles move faster, we see this as a rise in temperature.

Particle ideas can help pupils to picture what is happening in a change of state. As we increase the temperature of a solid, the particles move faster; their vibration increases. At some point, the vibration becomes so strong that the attractions of the neighbouring particles cannot hold the structure firmly together, and the substance becomes a liquid. If we increase the temperature further, the motion of the particles becomes so great that they break away completely from the attractive forces of their neighbours and the substance becomes a gas.

One way to help pupils visualise this process is to use a vibration generator, or an upturned loudspeaker cone connected to an audio-frequency signal generator, to make a collection of polystyrene balls vibrate. Initially the balls remain in a regular array. As the vibration increases, they become more disordered and eventually separate completely. This is, however, only a rather imperfect analogy. If you use it as a demonstration, it is important to stress that, unlike the polystyrene balls, the particles of a substance are in continuous motion because of their temperature and do not need any outside source to keep them moving. Another practical difficulty of this demonstration is that it can be hard to talk over the noise it makes! For these reasons you might prefer to use a computer package which shows animations of particle motion in solids, liquids and gases, as the temperature is changed. One that is widely used is the CD-ROM *States of Matter* (from New Media); this is available in two versions, one designed for pupil self-study and the other as a 'teaching tool' for use in class discussion.

Transfers of heat

A key element of the model of Figure 1.4 is the idea that heat moves spontaneously from regions of higher temperature to regions of lower temperature. At lower secondary stage, we can explore this further by looking at the different transfer mechanisms: *conduction*, *convection* and *radiation*. Textbooks for lower secondary stage describe a number of well known practical activities which can be used to illustrate aspects of these mechanisms. Your choice will depend on what equipment is available in your department. Many are best done as teacher demonstrations. Some can be used in a circus of short practical exercises, though one constraint is that many have to be allowed to cool down before they can be re-used by the next group. As these practical illustrations are well documented in many other texts, they will not be discussed in detail here.

One note of caution, however, is needed here. The thermal conductivities of metals are often compared by heating the ends of rods of different metals and observing the melting of wax at points along the rods. Whilst this is a useful demonstration, you should bear in mind that the results also depend on the heat capacities of the rods. A metal with a high heat capacity will appear to conduct more slowly than one with a low heat capacity.

When investigating radiation, never use mains-powered radiators with bare elements, even those protected with a fine mesh screen. When detecting radiation from a hot source using the back of the hand, take great care to avoid the danger of skin burns.

Thermal insulation

A very widely used practical task is to monitor the temperature of hot water in a set of beakers (or other containers), each wrapped in a different insulating material (Figure 1.7). Pupils should observe that a wrapped beaker cools more slowly than an unwrapped one, and that some wrapping materials work better than others. Some may have done this, in a simple form, at primary school. At lower secondary stage, they should learn to *explain* their observations using the model of Figure 1.4. We are, in effect, adding to the basic model the idea that the material between the two objects influences the rate at which heat is transferred.

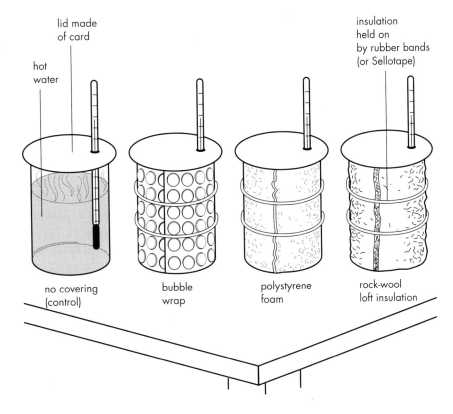

Figure 1.7
Investigating how the rate of cooling is affected by insulating materials.

lid made of card

insulation held on by rubber bands (or Sellotape)

hot water

no covering (control)

bubble wrap

polystyrene foam

rock-wool loft insulation

This may seem a very straightforward practical activity, but it raises a number of teaching points.

- Some pupils find *rate of cooling* a difficult variable to handle. Many find it easier to think about the temperature drop in a 10 minute interval, rather than collecting temperature readings at 1 minute intervals and plotting a temperature–time graph.
- It is difficult to make the initial temperature exactly the same in several containers. Hot water from a kettle is cooled by contact with the cold container, so even if freshly boiled water is used, its temperature may be below 90°C by the time the first measurement is made. It is worth pre-heating the containers with some hot water. Alternatively, wait until the hot water reaches a temperature chosen in advance, say 85°C, before beginning to monitor. This makes it harder, however, to collect results from several containers cooling together, as readings have to be taken at different times. Also, some pupils may think that to treat all the containers 'fairly' you have to start measurements as soon as the hot water has been added!
- Some pupils at lower secondary stage may think that good insulating materials are 'naturally warm' materials. If so, they may predict that materials which are best for keeping a hot drink hot would be worst for keeping a cold drink cold! This is a good question to ask to probe pupils' ideas.

'Thermal' processes

Those who think we should not use the word 'heat' as a noun may prefer to talk about the *'thermal transfer* of energy' in discussing the ideas above, rather than 'heat transfer'. It does not, however, make sense to talk of the 'transfer of *thermal energy*': the adjective 'thermal' describes the form of transfer, not the form of energy. We gain nothing – and probably cause additional confusion – if we simply replace the term 'heat' (which is widely used by ordinary people and by most scientists and engineers) by the term 'thermal energy' (which is not widely used by anyone!). But we *can* talk of 'thermal transfers' of energy – 'thermal' meaning that the transfer of energy results from a temperature difference (or, in particle terms, as a result of the random motion of atoms or molecules).

Heat, work and energy

The model of thermal processes in Figure 1.4, and some of the practical activities discussed previously, might seem to imply that 'heat is conserved' in thermal processes. This is incorrect. It was, however, the basis of the scientific understanding of thermal processes through much of the late 18th and early 19th centuries, and it is still a useful model for thinking about many everyday thermal processes.

The experimental work which convinced the scientific community that this idea was incorrect was carried out by James Joule during the 1840s; it involved very careful measurements of the temperature rise in samples of water, and other liquids, when agitated by paddle-wheels, turned by falling weights. Joule's results showed that a temperature rise could be produced by doing mechanical work on an object. His measurements required experimental skill of a high order and it is almost certainly not worth trying to replicate these investigations in a teaching situation.

There is, however, one point related to this with which pupils need to be familiar before you embark on the general treatment of energy ideas discussed in Section 1.3 – that objects can be heated by rubbing them. Most are likely to know this by lower secondary stage. Point out to them the heating effect of rubbing your hand on a bench. Or they could feel the brake blocks of a bicycle after they have been used several times to stop the wheel turning, or notice how a wooden block gets hotter when it is rubbed across another wooden surface for 20 seconds. The heating effect of friction can be shown more dramatically by using an electric drill to make several holes in a block of hard wood. After a short time, wisps of smoke can be seen coming from the wood block. This works best with a blunt drill bit.

! *You should wear eye protection and use a safety screen.*

What this shows us is that there are two ways (at least) of raising the temperature of an object:

- by heating (i.e. putting it in thermal contact with another object at a higher temperature);
- by doing mechanical work (when a force moves through a distance).

This means, of course, that we cannot use the word 'heat' both for the thing that is transferred to the object because of a temperature difference and for the thing that a hot object contains. In most situations they are not the same, and so we need two separate names for them. We can simply call the thing the hot object contains 'energy' (or 'internal energy'). We then use the word 'heating' for the *process* of energy transfer due to a temperature difference.

So, although it has been suggested that the word 'heat' should be used when we begin discussing thermal processes, so as to build upon the ideas and language which pupils will bring with them to the classroom, this is the point at which to move away from this everyday usage and adopt more consistent terminology.

Why drop the word 'heat'?

We *could* retain the word 'heat' as the name for energy transferred from one place to another because of a temperature difference. This is what 'heat' means in classical thermodynamics. We would, however, then describe a situation like that shown in Figure 1.4 by saying that some of the internal energy of the hot object is transferred 'as heat' to the colder object, where it becomes internal energy again. This seems unnecessarily complicated. It is simpler just to talk of energy being transferred from the hot object to the colder one. Pupils who choose to study physics at more advanced levels will, however, meet 'heat' with the meaning above, so for that reason you may prefer to continue to use it at this stage. The important thing is not to talk of heat as something that can *reside in* an object.

1.3 Energy transfers

Energy is an important idea in science because it provides a unifying framework for thinking about an enormously wide range of phenomena, which might otherwise appear completely unrelated. Having covered the ideas of Section 1.2, you are in a position to embark on a general discussion of energy and energy transfers. This is often part of a lower secondary science programme. The aim is to help pupils to see a wide variety of processes and events in which objects or materials change in some way, in terms of the storage, transfer, conservation and dissipation of energy.

One problem in introducing energy ideas is that it is very difficult to say clearly but simply what we mean by 'energy'. Here is how the idea was expressed by Richard Feynman, one of the 20th century's foremost physicists:

> 'There is a fact, or if you wish a law, governing all natural phenomena that are known to date. There is no exception to this law – it is exact so far as is known. The law is called conservation of energy. It says that there is a certain quantity, which we call energy, *that does not change* in the manifold changes which nature undergoes. That is a most *abstract* idea, because it is a mathematical principle; it says that there is a numerical quantity, which does not change when something happens. It is *not a description of a mechanism*, or anything concrete; it is just a strange fact that we can calculate some number and when we finish watching nature go through her tricks and calculate the number again, it is the same.'
>
> (R.P. Feynman, *The Feynman Lectures on Physics Volumes 1–3*)

In practice, we have to build upon the pupils' everyday understanding of the word 'energy' and try to refine this as we go along. So rather than trying to define energy, we can say that energy can be:

- stored (in various ways);
- transferred (from one place to another);

and that, in any process, energy is:

- conserved (that is, there is the same total amount of it at the end as there was at the beginning);
- dissipated (that is, it is more widely spread, or less 'concentrated', at the end than it was at the beginning).

We can also identify a number of ways in which energy can be stored:

- in a chemical substance, or group of substances, which act as a fuel (e.g. a hydrocarbon plus oxygen, or the chemicals inside a dry cell, or nuclear fuels such as enriched uranium);
- in a hot object;
- in a moving object;
- in an object that has been raised to a height (and, similarly, in a pair of attracting magnets or a pair of attracting electric charges that are separated from each other);
- in a springy object that has been stretched or compressed from its natural shape.

Every process or event starts with energy stored in one or more of these ways. At the end of the process, the energy will again be stored in one or more of these ways.

We may find it useful to use a shorthand label for each of these ways of storing energy, calling them, respectively, *chemical energy, internal energy, kinetic energy, field potential energy* and *elastic potential energy*. 'Electrical energy' and 'light energy' have been omitted from this list intentionally because these refer to ways in which energy can be transferred but are not ways in which it can be stored. There is more discussion of this point later. The label 'field potential energy', rather than 'gravitational potential energy', has been used so as to include situations involving electrostatics and magnetism.

Energy forms

Some people argue that labels for different forms of energy are unnecessary and unhelpful, as they may make pupils think that the different forms of energy really are different things. If so, it is possible to avoid forms and talk simply about 'energy', perhaps adding a few words to make clear whether it is associated with an object's movement, or position, or distortion of shape, or temperature. However, most physicists and engineers talk about kinetic energy, potential energy, and so on. Ideas about energy being transformed from one form to another also played a significant part in the thinking of some of the scientists who first struggled towards the notion of the conservation of energy, particularly Mayer and

Helmholtz. Whilst the most effective route for learners does not necessarily follow the original historical route to the ideas, the underlying notion of something remaining constant in quantity during changes, despite all the observable differences, seems important in coming to terms with the energy idea.

If you do choose to use labels for forms of energy, it is important to make clear that they are just convenient labels and that energy is the same thing, in whatever way it is stored. At a later stage, we may want to show how all of these can be reduced to kinetic energy or field potential energy, or a mixture of the two, by considering what is going on at the atomic or molecular level. However, when we move towards a quantitative treatment of energy, we will find that there are different formulae for calculating amounts of energy stored in each of the five ways identified above. So the choice of these five labels is not arbitrary.

People sometimes use the term 'stored energy' as a synonym for 'potential energy'. This may, however, be confusing for pupils, as all five of the forms above can be stored in an object or system of objects.

It is also useful to distinguish a number of distinct methods by which energy can be transferred from place to place:

- by *mechanical work* (when a force moves something through a distance);
- by *heating* (when a temperature difference causes a transfer of energy from one place to another by conduction or convection);
- by *radiation* (transfer of energy by electromagnetic radiation, or by mechanical waves such as sound);
- by *electrical work* (when an electric charge moves through a potential difference – this is the main method of energy transfer in electric circuits).

This set of ideas provides a framework which we can apply to a wide range of events and processes. The purpose of the outline above is to provide a concise summary of the ideas we want to teach. You would not, of course, introduce these ideas at the start, but build them up gradually using examples. A teaching sequence for doing this is outlined on the pages that follow.

◆ *Previous knowledge and experience*

Most pupils will not have met the idea of energy in their science programme at primary school. The Science National Curriculum for England and Wales does not mention 'energy' at Key Stages 1 and 2. Children at this age, however, are building up their knowledge of the natural world through informal everyday experiences. This provides a 'data base' which is essential for developing an understanding of the scientific idea of energy at secondary level.

◆ *A teaching sequence*

Looking at events from an energy point of view

A good way to begin is with a class discussion. You might show pupils some objects like a piece of coal or solid fuel, a small camping gas cylinder, a dry cell (battery), and say that these can be thought of as 'energy stores'. Based on their knowledge of fuels, and the way the word 'energy' is used in everyday conversation, most are likely to agree. Then you could ask how they know that there is energy stored in each of them. This is likely to lead to talk about transformations, in which the stored energy is transferred to other places and/or stored in other ways. You might then extend the idea of 'energy stores' to include a compressed (or wound up) spring and a raised object; and you might ask where the energy is stored at the end of transformation processes, pointing out that often something has become hotter.

Once some basic ideas about energy storage and transfer have been aired, you might focus the discussion on a few simple energy transfers which you can demonstrate and discuss. A good one to begin with is a battery running a small motor which raises a load. Choose a load that the motor can lift, but heavy enough not to accelerate up too rapidly! Start by asking pupils to think where the energy is stored at the beginning (in the battery) and at the end (in the raised load). Then ask them to think about the ways in which the energy is transferred: by electricity from the battery to the motor, and mechanically (by a force) from the motor to the load. It is then useful to summarise the process diagrammatically, as in Figure 1.8.

Figure 1.8
Energy transfers when a load is lifted.

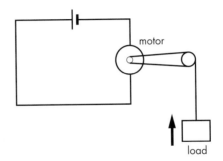

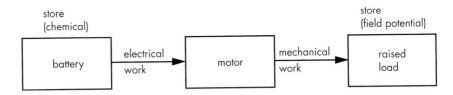

Another good introductory example is a simple catapult to make something move across the desk. Choose an object that rolls freely, so that you can concentrate on the transfer of energy from the stretched elastic to the moving object, without extending this (yet) to consider what happens next. An energy transfer diagram for this is shown in Figure 1.9.

Figure 1.9
Energy transfers using a catapult.

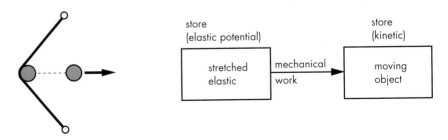

As a third example, you might ask pupils to consider a match being struck and burning till it goes out. This brings in the idea of the environment as the 'place' where the energy is finally stored, and the two other main transfer mechanisms, heating and radiation (Figure 1.10, overleaf). Radiation here includes both visible light and infra-red. It is also good to mention here that the original energy store is not just the chemicals in the match head but the *system* consisting of these chemicals plus oxygen from the air. By discussing a series of examples like these, you can introduce all the forms of energy storage and all the methods of energy transfer.

Figure 1.10
*Energy transfers
when a match
burns.*

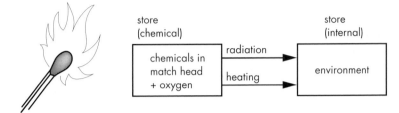

Energy transfer diagrams

In the box-and-arrow diagrams of Figures 1.8–1.10, each box
represents either a place where energy is stored or a device which
changes the method of energy transfer. The arrows indicate the
different methods of energy transfer involved. These diagrams are
simplified, focusing on the main energy changes only. In the first
two examples, some energy is also transferred to the environment
as a result of friction. We can extend the diagrams to indicate this
(Figure 1.11).

Figure 1.11
*Energy transfers,
environment
included.*

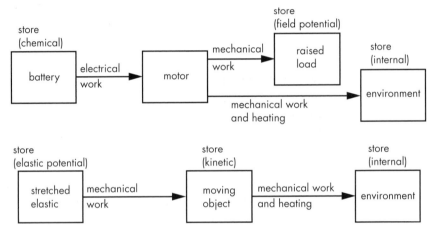

A second type of energy transfer diagram is the branching
arrow (or Sankey) diagram. Figure 1.12 shows Sankey
diagrams for the same two situations. An advantage of Sankey
diagrams is that the width of the arrows gives an indication of
the relative amounts of energy transferred in different ways
(although, of course, there is no way for the pupils to know
these relative amounts at this stage other than by being told).
The diagrams also imply conservation of energy. On the other
hand, the box-and-arrow diagrams indicate more clearly the
different methods of energy transfer involved.

No type of diagram can encapsulate *all* the features of the
energy transfer; they only summarise some key points. You will
find other kinds of diagrams in textbooks to represent energy
transfer processes, but it is better if you stick with just one or
both of these two types, and use them consistently.

Figure 1.12
*Sankey diagrams
can show relative
amounts of
energy
transferred.*

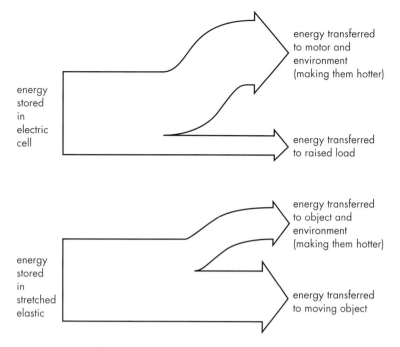

energy
stored
in
electric
cell

energy transferred
to motor and
environment
(making them hotter)

energy transferred
to raised load

energy
stored
in
stretched
elastic

energy transferred
to object and
environment
(making them hotter)

energy transferred
to moving object

An 'energy circus'

Pupils will now need time to come to terms with all these new
ideas. You might have them work in groups on a 'circus' of
small practical tasks, in which they observe a series of simple
energy transfer processes and try to summarise the energy
transfers by drawing box-and-arrow and/or Sankey diagrams.

The choice of examples for an energy circus is very
important. Ideally the set should include examples of all the
different ways in which energy can be stored and all the
methods of energy transfer. Some good ones to use are:

- a battery lighting a torch bulb;
- one of those small toys with a spring and a rubber sucker
  which can be stuck to a desk and which, after a short delay,
  jumps into the air – ask the pupils to think about the process
  up to the moment where the toy is at its highest point;
- a marble rolling down a ramp;
- a small piece of firelighter being used to heat some water in
  a beaker;
- a syringe with some air sealed inside, placed in a beaker of
  hot water – pupils observe the piston move;
- a small solar cell running a small motor and raising a small
  weight;
- a person running up a flight of stairs, or using an 'exercise
  step'.

Others could be added to this. For instance, the first example could work equally well with a buzzer, with energy being transferred to the environment by mechanical radiation (sound) rather than electromagnetic radiation (light). Or you could reverse one of the introductory demonstrations suggested on page 24, by using a falling mass to turn the spindle of a dynamo and generate an electric current.

Some teachers like to use everyday devices in an energy circus. If you do, take care to choose examples where the energy transfer is clear, and which include a range of forms of energy storage and methods of energy transfer (and see *Examples to avoid*, page 30).

Transfer or transform?

The last box (page 22) discussed the issue of using forms of energy or just talking about 'energy'. If you avoid forms then you have to talk about energy being 'transferred' rather than 'transformed'. This, however, leads to some difficulties in talking clearly about situations like the marble rolling down the ramp. It really feels very unnatural to say that 'the marble's initial potential energy is transferred to kinetic energy'; here 'transformed', or even 'converted', seems to communicate the idea much more clearly. The term 'transferred' tends to imply that the energy is somewhere else, when in fact it is still stored in the same object, but in a different way. Indeed even if you prefer to avoid forms of energy, it seems clearer to say that 'the marble's initial energy due to its raised position is changed (or transformed) into energy of the moving marble', rather than to talk of its being 'transferred'.

Some teaching points to consider when working on an energy circus are outlined in the following paragraphs.

Energy and fuels; emphasising the *system*

It is easy to convey the impression that fuels 'contain energy', and this can lead to quite persistent misunderstandings. More advanced students, even science graduates, are often unclear as to how, or where, the energy is 'stored' in a fuel. Some have the wrong idea that energy is released when the chemical bonds are broken – almost seeing energy as a kind of invisible fluid which can leak out of the broken bonds! So it is important to think carefully about how you want to talk about these ideas.

A hydrocarbon fuel does not 'contain energy'. It is one part of a system (the other part being oxygen) which can react to produce carbon dioxide and water vapour. In this reaction, energy is transferred from the fuel–oxygen system to other places; e.g. it can be used to make something else hotter.

Energy is *not* released when bonds are broken. Rather, it always requires some energy to break a bond, as we are in effect pulling two charged objects apart against their electrostatic attraction. Overall, however, energy is released when a hydrocarbon is burned with oxygen because more energy is released when new bonds form to make the combustion products than is needed to break the fuel and oxygen molecules apart in the first place. As Figure 1.13 shows for the case of methane, it is the weaker bonds in the oxygen molecule that are the key to the overall transfer of energy to other places. A spark is needed to start the process because an initial energy input is required to break the bonds in the first few molecules. Thereafter the energy released in the reaction breaks the bonds in more molecules of methane and oxygen, keeping the reaction going.

Figure 1.13
Energy is released when chemical bonds are formed.
(Adapted from: There is no energy in food and fuels by K.A. Ross, School Science Review, Dec 1993, 75 (271), p.39.)

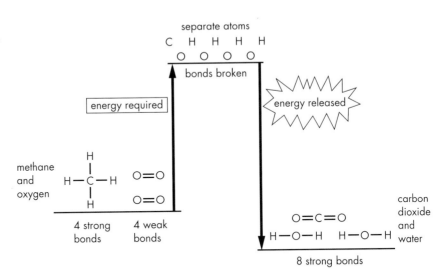

Although these ideas would not normally be introduced until upper secondary (or even post-16) level, and in chemistry rather than physics courses, it is still useful for you to think them through before talking to lower secondary pupils about energy transfers involving fuels.

'Missing' energy forms?

It was suggested earlier that it is better to avoid talking about 'light energy' and 'electrical energy'. The reason is that we cannot *store* energy in the form of light (or any other kind of radiation), nor as electricity. Radiation and electricity (or more precisely electrical work) are ways in which energy can be *transferred* from place to place.

Radiation transfers energy from a source to an absorber. Similarly, in an electric circuit, energy is transferred from the source (the battery, or the power station in the case of mains electricity) to the load (a device or appliance of some sort). You might argue that, in these situations, 'light energy' and 'electrical energy' are being 'produced' and 'consumed' at the same rate. But why bring them in at all? It is easier just to focus on the initial and final energy stores, and to see electric current and radiation as methods of transfer, rather than as forms of energy.

Examples to avoid

When choosing examples for an energy circus, it is wise to avoid situations that involve maintained steady motion, such as an electric drill or a car travelling along a level road at constant velocity. The reason is that there is a temptation, because the purpose of these devices is to produce motion, to see them as involving a transfer of energy to a moving object, which stores kinetic energy. But once the steady speed has been reached, the amount of kinetic energy stored remains constant – even though energy continues to be transferred from the source. All of the energy being transferred from the initial energy store is going straight to the environment, making it hotter. From an energy point of view, the devices are just like heaters! Only during the short period when the drill is speeding up initially or when the car is accelerating, is some energy being transferred to the moving object and stored as kinetic energy. Situations of this sort can be very useful to discuss with some pupils later in a teaching programme, once they have grasped the basic ideas, but should be avoided in the early stages.

Making the boundaries clear

You will need to help pupils to focus on specific aspects of processes and not get lost in complexities. For example, if we are looking at a simple process such as lighting a lamp using a dry cell, then we want to take the chemicals in the cell as the starting point and not worry unduly about the energy involved in getting them and putting them together in a cell. If we are considering a process where the initial energy store is a hydrocarbon fuel plus oxygen, then we do not want (at this point) to be concerned about tracing the energy back to the Sun. Similarly, we may want to draw a boundary before the final endpoint of a process, as we did in the initial discussion of the examples in Figure 1.11, for instance.

'Causes' not 'triggers'

You may need to help some pupils distinguish between things that 'cause' an event, in the sense of being a 'trigger', and those that are part of the energy chain. For example, if you ask a class about energy transfers in a simple circuit of battery and lamp, some may want to include the kinetic energy of the switch being closed! They argue that the closing of the switch is what causes the lamp to light. Of course, in one sense, this is true – but it plays no part in the chain of energy changes from the cell to the lamp and on to the environment. You need to be alert to the possibility of misunderstandings of this kind, and ready to talk them through if they arise. If this example is included in an energy circus, it may be better to leave the lamp switched on, so as to avoid any issues associated with the closing of the switch.

Where is the energy stored?

Some pupils get confused about where the energy is stored when an object is raised to a height. We often talk of the energy being stored *in* the raised object. But of course it only has more energy at a height because it has been moved in the Earth's gravitational field. The energy is really stored in the *system* comprising the raised object and the Earth. We might say it is stored in the 'object-in-the-field'. Pupils may find it easier to think about this if you also use examples involving magnetic and electrostatic fields. If we separate two attracting magnets by a short distance and then release them, they will move back together. When they are apart, energy is stored in the position of the magnets in the magnetic field. The same is true of two attracting charges which have been separated. All of these are examples of field potential energy.

Friction

Figure 1.11 indicated how the transfer of energy to the environment by friction, resulting in a rise in temperature, can be shown in energy transfer diagrams. In fact the box-and-arrow diagrams in Figure 1.11 are simplified and it may be worth discussing the processes involved in more detail with some pupils. First energy is transferred by mechanical work (by the frictional forces) to the surfaces which are rubbing. If we allow a little more time to pass, these hotter regions then transfer energy by heating to other objects nearby. Figure 1.14 shows how we might represent more fully the energy changes when an object is catapulted across a desk and gradually slows down and stops.

Figure 1.14
Energy transfers when a catapulted object comes to rest.

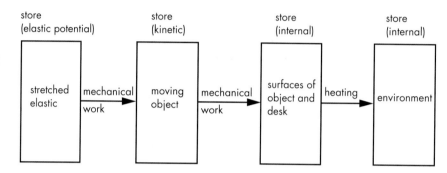

More complex energy transfers

Once pupils are able to apply ideas about energy transfer to simple situations, you might ask them to consider some more complicated examples, such as:

- the energy transfers from a coal- or gas-fired power station to a light bulb in their home;
- the manufacture of starch and glucose in the leaves of green plants.

At this point, you might also want to discuss how many energy sources can be traced back to the Sun as the original source. This can provide a link with the work on fuels in Section 1.1. You might discuss the energy transfer chain from the Sun to a renewable fuel such as wood, or to a fossil fuel like oil or gas, or to a device running on electricity generated by a wind turbine, or to our food.

Conservation and dissipation

Two key ideas about energy are that energy is conserved but is also dissipated in any process. That is, the total amount of energy remains constant, but the energy spreads out from being more concentrated to less concentrated, and so becomes less useful. It is important to introduce *both* these ideas *together.* The second is much more in line with our intuition that something is 'lost' or 'used up' in processes and can help resolve the contradiction which might otherwise appear between the need to 'save energy' and the fact that energy is always conserved. What we need to 'save' or 'conserve' (that is, to value and use carefully) are concentrated stores of energy. You should confront the two senses of the word 'save', and make clear how the scientific idea is compatible with everyday notions.

As mentioned above, Sankey diagrams (Figure 1.12) convey an implicit message about energy conservation: the total width of the arrows is the same on the input and output sides. That is, if we add up all the amounts of energy transferred in various ways to different places, this is equal to the amount of energy available at the beginning. You might want to introduce the joule as the unit of energy, even though you cannot yet explain how it is defined. Pupils may also have heard of 'calories' which you can explain are old-fashioned energy units, now superseded by the joule in science.

This is also a good point to discuss the idea of the *efficiency* of energy transfers (see the sub-section *Fuel efficiency* in Section 1.1). The key idea is that the energy transfers we want to achieve are always accompanied by other transfers that we do not want. Efficiency is a way of measuring how much energy goes where we want it to go. Again Sankey diagrams can help in getting this idea across. We can define:

$$\% \text{ efficiency} = \frac{\text{energy transferred to where you want it to go}}{\text{total energy transferred from source}} \times 100$$

The idea of conservation of energy, however, really only comes into its own when we move on to a quantitative treatment of energy at upper secondary stage and can calculate amounts of energy in different places. At lower secondary stage, we introduce the principle of conservation of energy but cannot provide any real evidence to support it. Instead we are simply aiming to help pupils understand what the idea means. Later we may be able to test it a little more rigorously, though the teaching laboratory is not a place where a fundamental principle like this can be 'tested experimentally'! Instead we accept it because it has proved useful to many scientists and engineers over a long period of time, by leading to reliable predictions in a wide range of practical situations.

Explanations at the particle level

It is possible to introduce these general ideas about energy transfers without mentioning the particulate nature of matter, but it can be helpful to pupils to make some links between the two areas. When discussing heat and temperature (Section 1.2) we introduced the idea that an increase in the temperature of a substance is, at the particle level, an increase in the movement of the particles. Now that you have introduced ideas like kinetic and potential energy, you can explore these links further. Some pupils find it helpful to think of the internal energy of an object as the total kinetic and field potential energy of all its particles.

This can also help pupils to think about what is happening in a change of state. You can explain that, if we heat a substance without changing its state, we are increasing the kinetic energy of its molecules. When we melt a solid or vaporise a liquid, however, we are separating or rearranging the molecules against their electrostatic attraction. This increases the field potential energy of each molecule, and so energy has to be transferred to the substance to make it happen. Because the kinetic energy of the molecules is not changing, the temperature of the substance stays the same during the change of state.

This can be observed practically, by using a boiling water bath to heat a sample of wax (or salol) in a boiling tube, and monitoring its temperature as it melts. The graph of temperature against time has a flat section at the melting point of the wax (Figure 1.15). During this period, energy is being steadily supplied to the wax. This is increasing the potential energy of the molecules of wax. Their kinetic energy stays the same, so the temperature stays constant. Although this is more often done as a cooling curve – allowing a molten substance to solidify and monitoring the temperature as it does – the 'heating curve' is easier to explain and easier for most pupils to understand.

Figure 1.15
Temperature stays constant as a pure substance melts.

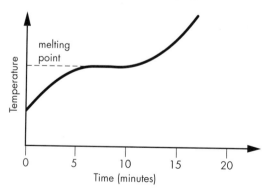

Particle ideas also help us to explain heating by friction. The rubbing of two surfaces makes some of the particles in the surface layer vibrate more rapidly, which results in a rise in temperature. This is then transferred (usually by conduction) to other parts of the objects. Similarly, we can see elastic potential energy as (electrostatic) field potential energy at the molecular level, and chemical energy as due to changes in (electrostatic) field potential energy within the molecules themselves. So all the forms of energy storage can be reduced to either kinetic energy or field potential energy, or a mixture of the two.

Energy and difference

Although an initial energy store is necessary for processes and events to occur, it is a mistake to say that energy *causes* things to happen. So, for example, we cannot say that petrol (plus oxygen) makes a car go *because* it contains energy. Energy is conserved in all processes, and as there is the same amount of it at the end as there was at the beginning, then it cannot have caused the process. If it did, then why does it not also cause the reverse process?

In fact changes always occur in the direction of increasing entropy, or decreasing 'free energy'. The total entropy of everything involved increases, and the total 'free energy' decreases. So 'free energy' might be said to cause processes to occur. But these ideas are too abstract and difficult for school physics courses. Recently, however, an approach has been developed and tested which uses the idea that *difference* is the cause of change. A difference in concentration or temperature can lead to spontaneous changes which reduce the difference; raising an object in a gravitational field (or separating attracting magnets or charges) causes a difference; changing a spring from its rest shape creates a difference; and the movement of a moving object makes it different from its surroundings. For more information on this approach, see *Energy and Change* (Boohan & Ogborn, 1996).

Power

Pupils may use the word 'power' when discussing energy transfers. In everyday language, the idea of power is not clearly differentiated from others like force and energy. You should take the opportunity, when the word arises, to explain its precise meaning in science as the *rate* of energy transfer.

Later, at upper secondary stage, when we begin to treat energy ideas more quantitatively, you can introduce the equation:

$$\text{power} = \frac{\text{energy transferred}}{\text{time}}$$

1.4 Measuring energy changes and transfers

At upper secondary level we can develop the ideas introduced in Sections 1.1 to 1.3 by considering how to measure amounts of energy transferred in different situations. It is only now that we are getting close to the scientific idea of energy as a quantity which can be calculated. The formal scientific definition of energy is 'the capacity of a body or system to do work, under ideal conditions'. In order to begin to understand this definition, it is necessary first to understand what is meant by 'work'.

♦ *A teaching sequence*

Measuring mechanical transfers of energy

A good way to begin a more quantitative treatment of energy ideas is with a class discussion of how you might get a stationary object moving, or raise an object to a height. The pupils will probably suggest methods that involve either a person or a device (such as a motor) as the active agent. You could then discuss how to summarise the energy transfer process diagrammatically, bringing out the idea that mechanical *work* is the method by which energy is transferred to the object (Figure 1.16). This then leads on to the idea that, if we could measure the amount of mechanical work done, we would know how much more kinetic or potential energy the object now had (provided it is reasonable to assume that there is little dissipation, so that all the energy transferred is now stored in these forms).

Figure 1.16
Work as a way of transferring energy: lifting a load.

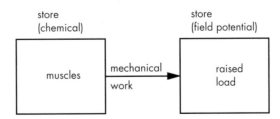

Work is done when a force makes something move, so you can propose a way of measuring amounts of work:

$$\text{work done by a force} = \text{force} \times \frac{\text{distance moved in the}}{\text{direction of the force}}$$

Most pupils find this definition intuitively reasonable. You can explain that the unit of work will therefore be the newton-metre, which is more commonly called the joule (J). As mechanical work is a means of transferring energy, it then follows that energy, too, is measured in joules.

You might then ask pupils to carry out a number of lifting, pushing and pulling tasks around the laboratory, using a forcemeter to measure the force involved and a ruler to measure the distance through which the force moves. By calculating the work done in each case, they gain familiarity with the idea of work and with the equation for calculating amounts of it. They also get a feel for how big or small a joule is. A useful snippet of information for pupils to remember is that it takes a force of about 1 N to lift an apple, and about 1 J of energy is transferred when you lift an apple from the floor on to a table.

It is important to point out to pupils that the meaning of 'work' in physics is *not* the same as its everyday meaning. In physics, no work is done unless a force makes the object move. So we do no work when holding a heavy box or suitcase, only when we raise it. To make this seem reasonable, you could point out that a shelf does not need fuel to enable it to keep holding things up! The reason why holding a heavy weight feels like work is to do with how our muscles act – by constant small contractions and relaxations. So in fact, there *is* work (in the physics sense) being done inside the muscles, even though the weight is not moving, and that is why it feels like 'doing work'!

Measuring gravitational potential energy

From the equation for calculating work, it is straightforward to derive an equation for calculating amounts of field potential energy in the Earth's gravitational field (or gravitational potential energy, for short). You might lead the class through the argument. Start from an energy transfer diagram (Figure 1.16) for a load being lifted. Energy is being transferred by mechanical work, so the increase in the amount of gravitational potential energy stored will be equal to the amount of mechanical work done, assuming that there is no significant transfer of energy to the surroundings by heating, due to friction. That is:

gain in gravitational potential energy = weight × gain in height

or

$$\Delta E_p = mg \times \Delta h$$

Here m is the mass of the object, g the strength of the Earth's gravitational field (in N/kg) and Δh is the gain in height. The deltas (Δ) emphasise that we can only calculate *changes* in gravitational potential energy. The baseline we choose as the zero level is entirely arbitrary.

An enjoyable activity is to ask the pupils to work out the amount of stored chemical energy they convert to gravitational potential energy each time they stand up from a squatting position, by measuring their weight and estimating the distance their centre of mass rises. They might also think about where the energy has gone if they now squat back down again! You might ask them to calculate how many knee-bends are equivalent to the energy stored in a Mars bar (according to the wrapper). This will be a rather large number – which allows you to bring out some ideas about the efficiency of the energy transfer processes involved. Most of the energy ends up making you feel hotter!

Measuring kinetic energy

In pre-16 science courses it is usual simply to state the formula for kinetic energy rather than try to derive it:

kinetic energy of a moving body $= \frac{1}{2}mv^2$

Here m is the mass of the body and v is its velocity.

Energy transfers between gravitational potential energy and kinetic energy

A nice demonstration at this point is the pin and pendulum (Figure 1.17). The first step is to make sure pupils know that a pendulum bob will rise again to its initial starting height after a swing. Then ask them to predict what will happen if the string hits a pin fixed horizontally, directly below the point of suspension. The bob still comes up to its initial height.

Figure 1.17

Interrupting the swing of a pendulum by a horizontal pin.

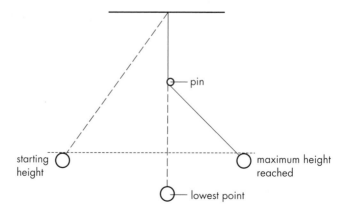

Using the equations for kinetic energy and gravitational potential energy, you can show pupils some of the advantages of the energy approach. For example, we can predict the speed an object will reach by sliding down a smooth slope *of any profile*. The shape of the slope doesn't matter. It is much easier, however, to talk about this as a 'thought experiment' than to try to show it practically! You should do some simple calculations of the speed reached by objects sliding down smooth ramps. One point that emerges is that the mass of the object doesn't matter – although many pupils will find it easier to do the calculations if you supply this information initially, so that the calculation can be done by arithmetic rather than algebra. Later, with some, you might simply call the mass *m*. It also makes the calculation easier if you choose a numerical value for the change in height which makes the speed come out as an integer (such as 1.25 m, 5 m or 20 m, with $g = 10$ N/kg).

Another useful exercise involves measuring the final speed of a dynamics trolley pulled for a certain distance by a falling mass. Attach the mass to a trolley using string, and hang it over a pulley at the end of the runway. The gravitational potential energy lost by the falling mass can be compared with the gain in kinetic energy of the moving trolley and the falling mass. It is important to use a free-running trolley and a well-oiled pulley, otherwise a lot of energy will be transferred to parts of the trolley and pulley that get hotter – changing the expected result.

Machines and efficiency

At this point, pupils might investigate the efficiency of a simple mechanical machine as an energy transfer device. A good one to use is a pulley system, as this seems more obviously a 'machine' than does a ramp or a lever. Figure 1.18 (overleaf) shows an arrangement using two double pulleys, although you could use 1:1 or 2:1 arrangements, or triple pulleys, depending on what you have available. Make sure they are well-oiled and free-running. The task for the pupils is to measure the force needed to raise the load by a measured distance, and the distance the applied force has to move to achieve this. Many pupils find it interesting that the applied force is less than the weight of the load.

Figure 1.18
*Using a pulley
system to
investigate
efficiency of
energy transfers.*

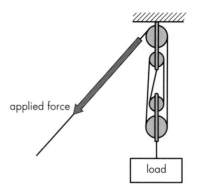

applied force

load

They can then calculate the gain in gravitational potential of the load and compare this with the work done by the applied force. The latter will be larger. This is not too difficult to explain; there is a fair amount of friction in a pulley system, and the lower pulley has been raised, as well as the load. From their data, they can calculate the percentage efficiency.

$$\% \text{ efficiency} = \frac{\text{energy transferred to where you want it to go}}{\text{total energy transferred from source}} \times 100$$

So, in this situation:

$$\% \text{ efficiency} = \frac{\text{gain in gravitational potential energy of the load}}{\text{work done by the applied force}} \times 100$$

As an extension, you might ask pupils to compare the efficiency of different pulley systems.

Measuring changes of internal energy

A good practical activity, which could be developed as an investigation, is to explore the effects of supplying different amounts of energy to different objects and noting the temperature rise produced. The easiest way to do this is with an electrical immersion heater and samples of water.

Wear safety glasses. Heaters with cracked seals have been known to explode.

There is no need for any electrical measurements – just use the same heater and vary the time for which it is switched on. The amount of energy supplied is proportional to the time. It is fairly easy to show that the temperature rise is proportional to the amount of energy supplied when you use samples of the same mass. By using samples of different masses, you can show that the temperature rise gets less as the mass increases. With some students, you might want to point out that they are inversely proportional.

To get convincing results in all these practical exercises, you need to use well-insulated containers which do not, themselves, require much energy to heat them up (i.e. which have a low heat capacity). It is also better to supply energy for a relatively short time, enough for the temperature rise to be easily measurable but not too large. If you heat for too long, energy losses to the surroundings make it harder to see the pattern in the results.

With the same apparatus, it is also easy to show that different materials increase in temperature by different amounts when the same amount of energy is transferred to them. You could compare water, sunflower oil and white spirit.

You can then bring these ideas together in the equation:

$$\text{change in internal energy of an object} = cm \times \Delta T$$

Here, m is the mass of material and ΔT is the temperature rise. The usual name for the constant c, which depends on the material being heated, is *specific heat capacity*. If you want to avoid using the word 'heat', then perhaps the best solution is just to call it s.h.c.

Changes of state
The energy that you have to transfer to an object to cause a change of state (and which is released again when the change of state is reversed) is sometimes called 'latent heat'. The name dates back to the 18th century; the heat was latent (meaning hidden) because it did not show up as a rise in temperature. The amount of energy that has to be transferred to melt or to vaporise a sample of a substance depends on the substance and its mass. Although the ideas involved are straightforward enough, latent heat is not on most physics syllabuses until the post-16 stage, so it will not be discussed further here.

Measuring energy transfers in electric circuits
It is useful to introduce energy ideas at an early stage in the discussion of electric circuits, as this can help pupils to reconcile the fact that electric current is the same at all points around a series circuit with their intuitive knowledge that batteries run down and so 'something' must be used up. To make sense of this, pupils need to separate the idea of energy (which is transferred from the battery to the device and on to the environment, and does not come back) from current (which circulates around the circuit and is the means by which the energy is transferred). Even though it is not strictly correct, it can be helpful to talk of the moving charges as 'carriers' of the energy. These ideas are likely to be introduced at lower secondary stage.

Later, at upper secondary stage, we can develop the idea of potential difference as a measure of the energy transferred between two points by each unit of charge which passes:

$$\text{potential difference} = \frac{\text{energy transferred}}{\text{charge}}$$

The unit of potential difference is the volt, which is therefore equivalent to the joule per coulomb. From this, we can then quickly derive the result that, in an electric circuit:

energy transferred per second between two points = VI

Here V is the potential difference between the two points in question, and I is the current. Pupils will then need practice in using this equation to calculate the amount of energy transferred by electrical work in a variety of situations. You might want to link these ideas to domestic electricity use, and to other ideas outlined earlier in Section 1.1. For example, pupils might use the power ratings (in watts) of domestic appliances to work out the amount of energy transferred in given periods of time, and relate this to the cost. This can highlight the point that devices used for heating are expensive in terms of energy (and cash).

◆ *References*

Boohan, Richard, and Ogborn, Jon, 1996: *Energy and Change.* ASE publications.

Centre for Alternative Technology: *Teachers' Guides: Wind Power; Solar Heating.* Centre for Alternative Technology, Machynlleth, Powys, SY20 9AZ.

Hunt, Andrew, and Milner, Bryan (general eds), 1992: *Pathways Through Science: Energy Resources.* Addison Wesley Longman Schoolbooks.

 New Media: *States of Matter* (CD-ROM). New Media, P.O. Box 4441, Henley-on-Thames, Oxon, RG9 3YR.

◆ *Other resources*

◆ The *SATIS* materials (ASE, 1986–) are a useful source of ideas for activities related to fuels. Among the units which you might use here are: **106** The design game, **107** Ashton Island – A problem in renewable energy, **109** Nuclear power, **201** Energy from biomass, **308** The second law of – what?, **403** Britain's energy sources, **601** Electricity on demand, and **908** Why not combined heat and power? There are also several useful sections in the *SATIS Atlas*.

- The *SATIS 16–19* materials (ASE, 1990–) also contain some useful units: **21** Energy from the wind, **46** Energy from the waves, **63** Biogas, and **96** Barrage report, although these would need to be modified for use with younger pupils.
- Materials for a role-play exercise on siting a windfarm are provided in *Teaching and Learning about the Environment Pack 3* (UYSEG, 1991), available from ASE Booksales.

Web sites

There are many web sites with information on fuels and energy. Some of the most useful ones are given below.

The Department of Trade and Industry provides data on fuel use in the UK:
www.dti.gov.uk/public/expl.html
(Choose the *Activities & Resources* option, then *Energy Statistics*, and then *Energy in Brief*.)

A useful site at the University of Oregon has good background information on renewable energy sources, and many links to other energy sites worldwide:
www.zebu.uoregon.edu/energy.html

The Energy Technology Support Unit (ETSU) has links to many other sites on energy topics:
www.etsu.com/home.html

The Centre for Alternative Technology provides useful introductory information on renewable sources:
www.cat.org

Background reading

For general background information on fuels and fuel use, together with simple introductory accounts of the technology of the various renewable energy sources, and some discussion of pros and cons:

Elliott, David, 1997: *Energy, Society and Environment.* Routledge.
Ramage, Janet, 1997: *Energy: A Guidebook* 2nd edition. Oxford University Press.

For a little historical background to how the scientific idea of energy was developed, try Chapters 4 and 5 of:
Spielberg, Nathan, and Anderson, Byron D., 1987: *Seven Ideas that Shook the Universe.* John Wiley.

2 *Sound, light and waves*

Martin Hollins

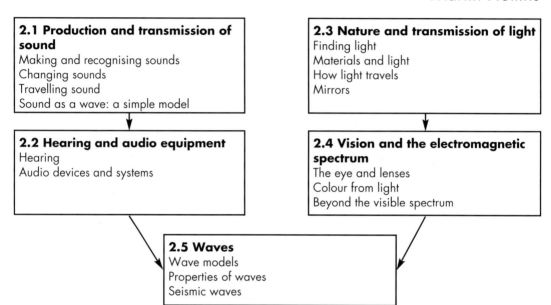

2.1 Production and transmission of sound
Making and recognising sounds
Changing sounds
Travelling sound
Sound as a wave: a simple model

2.2 Hearing and audio equipment
Hearing
Audio devices and systems

2.3 Nature and transmission of light
Finding light
Materials and light
How light travels
Mirrors

2.4 Vision and the electromagnetic spectrum
The eye and lenses
Colour from light
Beyond the visible spectrum

2.5 Waves
Wave models
Properties of waves
Seismic waves

◆ *Choosing a route*

Sound and light will have been studied in some detail by all pupils in primary school. They will probably have enjoyed much of that work and have some clear understanding of parts of the topics. Sections 2.1 and 2.3 are suitable starting points for lower secondary pupils. In each of these, details are given of pupils' likely experience in primary science lessons, and of the probable extent of their understanding. Sections 2.2 and 2.4 build on these and are suitable for later study.

Sound is a relatively accessible subject, and one where pupils will see many applications to everyday life. Light can be a challenge to pupils' understanding; much of optics, which used to be included at this level, has little direct appeal. By contrast, the uses of both sound/ultrasound and light/electromagnetic waves can provide attractive areas of enquiry for pupils in upper secondary classes.

The wave concept (Section 2.5) is a powerful model for explaining the behaviour and similarities of sound and light. It provides a good climax to this sequence, but it is complex. It is recommended that the topic is covered as late as possible; much of sound and light can be taught without reference to waves.

2.1 Production and transmission of sound

♦ *Previous knowledge and experience*

Through the primary phase children will have been developing their ideas of what makes a sound and how it travels. They will have a broad experience of making sounds (musical and otherwise!) and will have been encouraged to explore pitch and loudness. They will have been introduced to the concept of vibration and how this can be used to explain variations in pitch and loudness. They will have experience of sound travelling through a range of media, including solids and liquids, and have looked at the ear as a receiver of sound. There is a video, *Making Sense of Science: Sound*, which shows a range of good primary practice on the topic (SPE, 1996).

Children's ideas about sound have been investigated in some detail and incorporated in some primary teaching resources (see *Other resources* at the end of the chapter). The challenge for the teacher is to move pupils on from associating sound with properties of materials and human action to a confident grasp of the vibration concept. This will involve providing examples of progressively less obvious vibrating sources. At 4 years old children notice the vibration of a cymbal, but even at age 16 pupils may have difficulty with the idea of transferring the vibration of the object to air. Suitable strategies include:

- devising ways of making the imperceptible perceptible (e.g. sand or rice on a drum skin to show vibration);
- giving pupils the opportunity to investigate their own ideas (e.g. they may think that the speed of sound depends on its pitch);
- encouraging pupils to generalise their findings (e.g. about the size of a musical instrument and the pitch of the sounds it makes);
- encouraging pupils to be more specific in their use of words (e.g. distinguish between low for pitch and soft for volume or loudness).

◆ *A teaching sequence*

This is a good topic for developing positive attitudes to science. Most pupils are keenly interested in music and many will play instruments. There are opportunities for addressing personal and social interests such as how music is created and insulating against unwanted sound. There is scope for an investigative approach to the key concepts of how different sounds are made and transmitted.

It is inevitable that pupils will have met some of these ideas and activities previously. You may have to be prepared to move rapidly on if this becomes apparent. Some activities can be presented in terms of 'finding out what you already know'. You can maintain interest by emphasising some of the more technological aspects which are likely to be less familiar; for example, you could discuss voice recognition, synthesisers, sound meters.

Making and recognising sounds

Provide a range of musical instruments or home-made sound makers, or play a tape of familiar sounds. Ask pupils to describe the sounds and discuss with them their use of words such as 'low', 'high', etc.

Home-made musical instruments can include:

- flexible plastic and wooden rulers (take care they are not snapped by over-enthusiastic flexing);
- identical nails hammered into a block of wood to leave different lengths for striking;
- Savart's wheel – a disc with holes at different diameters from the centre which is rotated – blowing through different holes changes the note;
- rubber bands of different thickness and tension, stretched across an open box;
- straws of different length with a shaped reed at the mouthpiece (flatten about 1 cm length and cut each side at an angle of about 20°);
- bottles or boiling tubes filled with water to different depths.

Ask pupils to apply a simple classification scheme to how sounds are made, e.g. by percussion (striking a surface), string movement or wind (e.g. blowing across an aperture). Workcard pictures of musical instruments could be useful in organising this activity. Alternatively, ask pupils to bring in or describe instruments with which they are familiar, and try to incorporate these into the classification system. It would be good to include instruments from non-European cultures and to explore similarities and differences. Borrow a synthesiser from the music department and ask the class to 'guess the instrument'.

Demonstrate how a source of sound vibrates, using a tuning fork in water or against a ping pong ball, or put polystyrene beads into a loudspeaker cone (Figure 2.1). Ask pupils to 'hunt the vibration', identifying what it is that vibrates with other sources of sound in the classroom and elsewhere in their daily lives.

Figure 2.1
Showing sound vibrations.

A loudspeaker with an open cone is very useful for showing how electrical signals can be transformed into a sound vibration. The rear part houses magnets. Attached to the light, stiff cone of card is a coil which becomes an electromagnet when a current flows. It is attracted to the magnets which causes the cone to move. This movement may be visible and can be made more obvious with polystyrene beads.

47

Changing sounds

Demonstrate with a guitar, recorder and 'talking' drum (adjustable tension), or similar instruments, how notes of different *pitch* and *loudness* can be made. Provide home-made instruments (see page 47) and ask pupils to observe how they can make the notes change. Encourage them to generalise to rules such as:

- the greater the movement, the louder the sound;
- the larger the sound maker, the lower the sound, etc.

These rules could be reinforced by showing a mass suspended from a spring. Pulling down further makes the spring's oscillation bigger; replacing the mass with a larger one makes the oscillation slower.

Travelling sound

The variation of loudness with distance from a source can be explored in a number of ways. A context of interest to pupils could be to use a personal stereo (Walkman) as the source of sound. The volume could be varied and the distance measured at which the sound can just not be heard (in a range of directions). This could be linked with the concern about young people damaging their hearing by listening to their personal stereos with the volume too high. Ask pupils to set the volume to their chosen level for listening to their favourite music, and measure the distance at which others can hear the sound. Medical evidence is that if you can hear it 2 metres away, the listener is damaging their hearing! (British Tinnitus Association, 1997). A more quantitative investigation of this is possible if a sound level meter is available. At this stage it is not necessary to go into the details of the decibel scale, nor the inverse square law relationship between sound intensity and distance from the source. Meter readings should show a fall of 6 dB for each doubling of the distance from the source.

The effect of materials on the transmission of sound can be linked to ideas of noise reduction. Show the silencing of a bell in an evacuated bell jar, to demonstrate that sound cannot travel through a vacuum. ('In space no one can hear you scream!' as the movie blurb dramatically expressed it.) A vacuum vessel (strong glass bell jar) has a bell or buzzer placed inside and the pump is turned on. You can show the air coming out with a piece of paper blown by the exhaust of the exit pipe. As the sound disappears you can see that the bell hammer is still striking, or if a buzzer is used, that current is still showing on a meter.

Show how musical instruments use sound boxes to amplify their sources, for example demonstrate a musical box mechanism with and without a sound box (or in the hand and on a table). Alternatively strike a tuning fork on a bung and compare the loudness when it is held in the hand or put with its shaft on the table. These show how sound travels better in a solid than in air – because the vibrations are transmitted with less energy loss. Pupils can explore this by listening with their ear on the table which someone taps. They will probably have investigated string telephones in primary school but you could remind them of this and also get them to listen to a gently struck metal fork, spoon or coathanger which is hung from a string in contact with the outer ear (Figure 2.2) – the effect is a surprisingly loud and full sound. They can listen to sounds travelling in liquids with the aid of an underwater listening tube (Figure 2.3, overleaf) and be reminded of the plumbing gurgles you can hear if you submerge your head in the bath!

Figure 2.2
Listening to a coathanger!

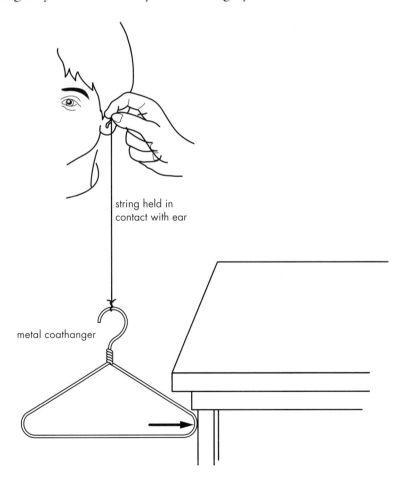

string held in contact with ear

metal coathanger

Figure 2.3
Listening under water.

balloon rubber
stretched over funnel

The speed of sound

The speed of sound in air can provide a challenge to pupils' measuring skills. It is possible to do it directly on the school field, with a suitable visible sound source (e.g. blocks of wood meeting) and with consideration for variables such as the wind. Alternatively pupils achieve an echo of a clap from the wall of a building, then clap immediately every time they hear the echo (i.e. in sync with the echoes) and measure the time for a series of claps and echoes. The value is about 330 m/s dependent on temperature (in practice, expect values within ±20% of this).

Typical results
time from first clap to eleventh clap = 9.5 s
∴ time for return journey to wall = 0.95 s
distance to wall = 140 m
∴ distance for return journey = 280 m
speed of sound = 280 m/0.95 s = 295 m/s

Ask your pupils whether they think sound travels faster in solids, liquids or gases. Most assume sound travels most easily in air, as it is 'thinner'. In fact, sound travels faster through solids. Some evidence for this can be gleaned from the traditional American Indian action of listening for horses' hooves by putting their ear to the ground. Pupils may have noticed that they can hear an approaching train first from the humming of the rails, before any sound arrives through the air. They should of course be discouraged from actually putting their ears to the rails – this is quite unnecessary!

Measurements of the speed of sound in the laboratory can be made using a dual beam oscilloscope with inputs directly from the source of sound and from the sound received at a measured distance (Figure 2.4). The oscilloscope traces enable the time difference to be measured accurately.

Figure 2.4
Measuring the speed of sound.

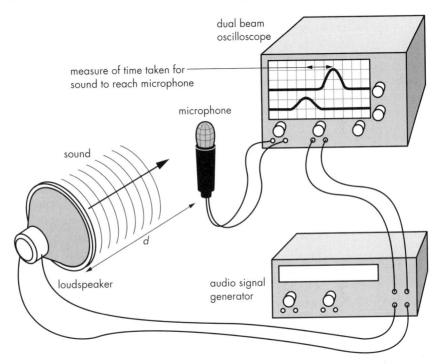

Reflected sound

Sound meeting an interface with another medium will be reflected or absorbed – usually both. The sound is reflected at an angle equal to that it approached with. This is the law of reflection. Hard surfaces tend to be good reflectors and smooth ones will reflect regularly. Pupils can investigate reflection using the arrangements in Figure 2.5 (overleaf).

Soft materials tend to absorb sound energy; pupils can investigate this by observing the loudness of a clock or personal stereo when placed in a box with different soft materials. A combination of reflectors and absorbers is often used to improve the acoustics of a room.

Figure 2.5

Investigating the reflection of sound.

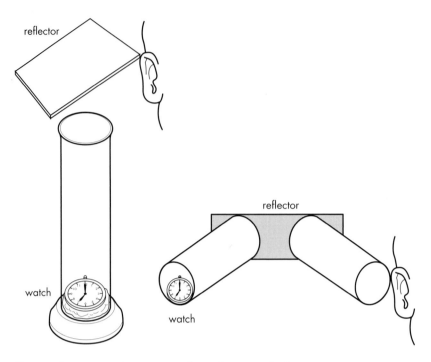

Sound as a wave: a simple model

The foregoing experiences can be linked together with a simple model of a sound wave which explains the role of the vibration essential to all sounds. The fuller picture of the wave model will be developed in Section 2.5. The display of a tuning fork making water ripples can be presented as a model of how sound itself spreads out like ripples over the surface of water. The ripples become shallower as they spread, because the energy is spread over a greater area. Making bigger ripples is like making a louder sound. The loudness still falls with distance, but the ripples will travel further before they disappear.

Displaying sounds on an oscilloscope is very useful at this point (see the *Equipment notes*, page 95). Alternatively, you could connect a microphone via a datalogger to a computer.

A microphone will show the shape from a tuning fork to be a sine wave; this is a pure note. Making the tuning fork sound louder increases the amplitude of the wave (Figure 2.6a). This is a good time to stress the link between the size of the vibration (amplitude) and the loudness of the sound, for example by reference to drum beats, etc. Putting the microphone to a fork with a higher pitch shows increased frequency and reduced wavelength (Figure 2.6b). Again reference can be made to familiar instruments for the link between faster vibration rates and higher pitches.

Figure 2.6
Oscilloscope displays of different sounds.
a The louder the sound, the greater the amplitude of the wave.
b The higher the pitch of the sound, the shorter the wavelength of the wave.

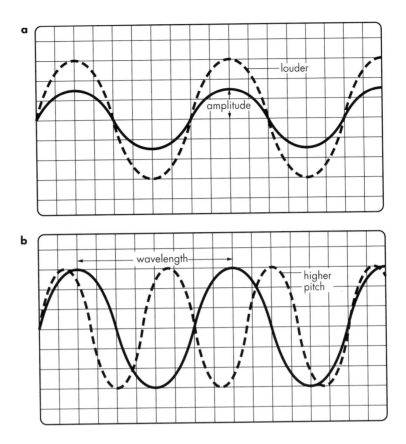

The same thing can be shown with a signal generator as the input to the oscilloscope (Figure 2.7, overleaf). Pupils enjoy trying this out and, of course, looking at the shape of sounds which they make themselves. Many will find it easier to whistle a pure note than to sing it!

A brief account of what is happening to the oscilloscope trace will suffice at this point, if pupils are not to be distracted by the details of the display system. Show how the timebase makes the spot move across the screen at different rates, and that the *y*-gain can be used to adjust the size of the vertical deflection. This enables sounds of different intensities to be seen. A microphone produces a voltage variation dependent on the intensity of the sound and the *y*-gain multiplies (or divides) this.

A detailed account of the longitudinal nature of sound waves is given in Section 2.5. At that stage, you will need to make sure that pupils have not gained (from what they see on the oscilloscope screen) the erroneous idea that sound waves are transverse.

Figure 2.7
*Oscilloscope set
up to display
sound waves.*

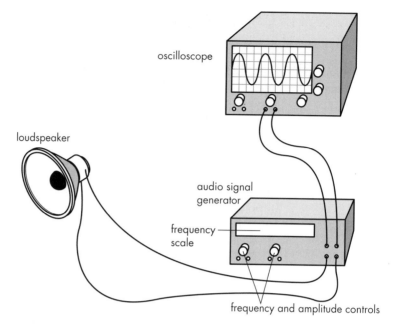

oscilloscope

loudspeaker

audio signal
generator

frequency
scale

frequency and amplitude controls

♦ *Further activities*

♦ A 'sound trail' around the school with a worksheet asking pupils to describe the sounds they hear, identify them and explain how they knew what they were.

♦ Measure sound levels around the school. Decide suitable levels for the library, private study areas, classrooms, the dining hall, etc.

♦ Design a new musical instrument. Predict how to make a range of notes and test the prediction.

♦ Design ear protectors for use in a noisy environment. Consider ear muffs, based on a headphone arrangement, and compare with ear plugs. Designs could incorporate plastic sponge, foam or other absorbent materials in the earpieces of audio headphones. Ear plugs are widely available (e.g. from Boots) as wax and as foam plastic versions, which could be compared. Less widely found are glass fibre ones, enclosed in plastic.

> **!** *Pupils should be warned never to stick anything sharp or thinner than their fingers into their ears because of the danger of rupture to the eardrum.*

♦ Compare the acoustics of two rooms, one with mainly hard surfaces, e.g. bathroom, and the other with soft furnishings, e.g. sitting room. Use the concept of reverberation time, and explain the different quality of the sounds heard.

◆ *Enhancement ideas*

- ◆ The original stethoscopes were simple rods placed between doctor's ear and patient's skin. Modern stethoscopes are tubes with special shaped bells to amplify the lower frequency sounds from the heart and the higher frequency sounds from the lungs.

- ◆ Water engineers listen with rods which they place on pipes where they think there is a leak. They hear turbulence where water is escaping. This can help them to detect people who are breaking hosepipe bans. Central heating bumps are transmitted efficiently and annoyingly around the house by the liquid and the metal pipes.

- ◆ Musical instruments which make sound mechanically (by hand) are called 'acoustic'. Instruments can be designed to use the mechanical motion to generate an electric current, as in the strings of an electric guitar. Electronic instruments such as keyboards do not need a mechanical vibration; pressing the keys closes switches so that oscillating currents flow.
  The varying currents from electric and electronic instruments are fed to an amplifier which increases their loudness (by increasing their amplitude), and can change the nature of the sound. 'Acoustic' sound can be converted with a microphone into electrical signals which can be amplified in the same way. Loudspeakers convert the energy carried by the electric current back into sound. Microphones and loudspeakers are examples of transducers; they transform one kind of energy into another. There is more about audio systems in Section 2.2.

- ◆ There is information on sound levels and recommended safe exposure times for loud sounds in Section 2.2.

- ◆ Sounds heard in a confined space have a very different quality from sounds heard out in the open. This is an effect of reverberation (repeated reflection). Reverberation can be added electronically. Great use is made of this in the soundtracks of films and radio plays.

2.2 Hearing and audio equipment

◆ *Previous knowledge and experience*

Hearing is an unproblematic concept for younger children, for example 'I heard ... because I listened' (Key Stage 1 child), and the 'active ear' concept is perhaps unwittingly encouraged by teachers' frequent injunction to pupils to 'listen (up)'. Progression requires the appreciation of sound as a vibration which is then transmitted through the ear and to the brain in various ways. There is a need for pupils to develop their rather general understanding of energy in order to consider the sensitivity of hearing and its possible damage.

The technology of music recording and playback can provide the context to associate the concepts of pitch and loudness with simple wave features. Pupils' widespread awareness of telecommunications probably hides a confusion about what exactly is transmitted. Some pupils for example think that TV pictures travel as light and the sound travels as sound waves. There are opportunities to demonstrate the differences of sound, light and radio transmission.

◆ *A teaching sequence*

Pupils' basic understanding of sound can be extended to a study of how and what we hear, possible defects of hearing and how to take care of hearing. Many pupils will be familiar with the use of audio equipment such as microphones and loudspeakers. Their study here can reinforce the concepts of sound as a vibration and the energy of sound. The use of ultrasound provides further interesting applications.

Hearing

Pupils might start this section of work with a period of silence – what can they hear? Does everyone hear the same things? If not, what reasons might there be for this? (Not just hearing sensitivity but also attention.) They could also explore the effect of *binaural* hearing – can sounds be located more easily with two ears than with only one? Are sounds heard equally well in all directions? Try sitting pupils in a circle, with one blindfolded at the centre. One pupil claps; the one in the centre has to indicate the direction of the sound. Repeat with one ear covered.

An audio signal generator (ASG) can be used to check the frequency range of the class's hearing – 'put your hand down when you fail to hear' – and the teacher will probably be first as this is quite age-dependent! Ask the class 'How could we tell if anyone is cheating?' (Turn off the ASG.) If the ASG is connected to an oscilloscope, the frequency change can be seen as well as heard. If the volume of the ASG output is varied then loudness can be seen to be associated with amplitude of the wave on the oscilloscope (see also Section 2.1 above). The range of hearing for an adult is typically from 20 Hz to 20 000 Hz (20 kHz). Pupils may hear as high as 30 kHz. You could explain that 'hertz' tells you the number of vibrations per second.

The ear

The ear is a complex organ and a full explanation is not required at this level. Textbooks will have suitable diagrams and teaching resources may include worksheets of ear parts to assemble (e.g. *Don't turn it off, turn it down*, British Tinnitus Association, 1997). A biological scale model is also useful for appreciating the relative sizes of the parts and how they are interconnected.

This should lead to an appreciation that the ear is a delicate organ and can suffer impairments. People have a range of permanent hearing loss, from mild and moderate for which hearing aids can compensate, to severe and profound. About a quarter of young people, it is estimated, listen to music loud enough to damage their hearing. This shows first as a temporary hearing loss or a 'ringing' noise called tinnitus. About 25% of these will go on to develop permanent tinnitus or hearing loss as they get older. The pack *Don't turn it off, turn it down* was developed to explore these issues with lower secondary pupils in science and PSE lessons.

Noise pollution

Noise can be seen as a kind of pollution – but what is 'noise'? Pupils could conduct a survey at school or at home to find out what sounds people most dislike. They could set up a database to analyse the results. Is the sound always loud or is some other property important? This will enable pupils to revisit the concepts of pitch and frequency, amplitude and loudness, and consider contexts (a sound may become 'noisy' after 11 p.m.!). Pupils could contact local environmental health officers to find out which noises are most complained about.

Acoustics

Acoustics is a topic which sets hearing in an interesting context, drawing attention to the properties of materials and shapes with respect to sound propagation. Pupils can investigate the sound absorbency of materials by putting a stop clock in a box lined with foam etc. Books and other resources can be consulted for details of sound absorbency in recording studios and acoustic design of concert halls.

Audio devices and systems

Making and using a loudspeaker and a microphone (the same device with input and output reversed) is a motivating activity for pupils which brings together several strands of learning – sound as vibrations, energy transfer and electromagnetism. The construction of a simple loudspeaker is shown in Figure 2.8. It requires a polystyrene cup, about 0.5 m of insulated single-strand copper wire, a magnadur or bar magnet, a low voltage/high current a.c. power supply or audio signal generator. Connecting it up should cause a vibration in the cup. Cling film may make this more obvious. Loudspeakers are very inefficient so don't expect a loud noise!

Then connect the device as input to an amplifier and loudspeaker. Speak into the cup or tap it to hear an output sound – again, the effect will be rather faint.

Figure 2.8
A model loudspeaker.

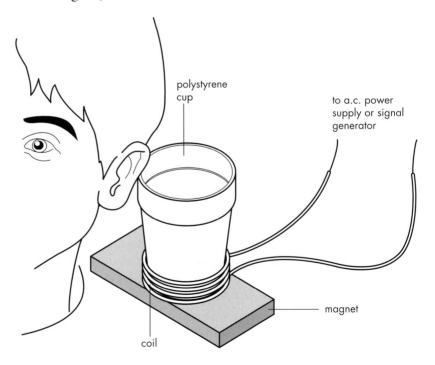

polystyrene cup

to a.c. power supply or signal generator

magnet

coil

An alternative is to investigate a telephone receiver. Mouthpiece/earpiece devices are often available from electronics shops and surplus stock agencies, and can be wired up for communication. Pupils should be reminded of the yoghurt pot telephone which they almost certainly will have built in primary school. This has no energy transformation – the sound is transmitted all the way by vibration.

A laser can be used to demonstrate light transmission of audio signals. A light strip of mirror is clipped or stuck to the loudspeaker cone so that it vibrates. When the laser beam is reflected from it and directed onto a detector such as a light-dependent resistor or a phototransistor, the sound can be reproduced through a second loudspeaker.

! *The laser must be a Class 2 laser – see the Equipment notes, page 95.*

Figure 2.9
Transmitting sound signals along an optical fibre.

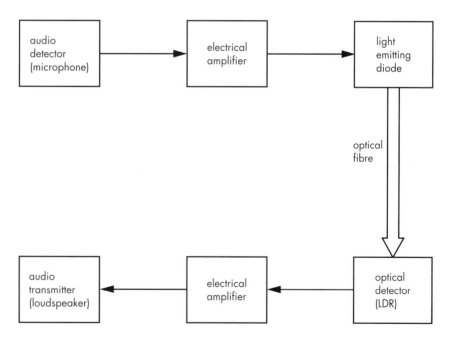

The signal generator can be replaced with a microphone and amplifier, or radio with socket for external speaker. Figure 2.9 shows a schematic diagram for setting up a similar demonstration using fibre optics (check equipment suppliers for suitable devices).

Discussion will probably extend to how mobile phones communicate. The similarities and differences between light and radio waves could be briefly mentioned.

The work on hearing will have established that the audible range for humans extends to a frequency of about 20 000 hertz (20 kHz). Higher frequencies are called *ultrasound*. An ASG can transmit at such a frequency and a microphone can receive it. If this is connected to an oscilloscope, pupils can see how the frequency of ultrasound compares with audible sound. Such frequencies are used in the animal world for communication. A dog whistle for example is inaudible to us. Bats and dolphins emit pulses of sound in the 30–100 kHz range and use the echoes to navigate. During the Second World War humans developed a way of using ultrasound echoes to locate underwater objects such as submarines. This system, SONAR, has developed considerably since then and is available cheaply to boating enthusiasts. Ultrasonics is now used for non-destructive testing in industry and for medical imaging. Unfortunately there is little that can be attempted practically in school, but ultrasound images are readily available and a visit to a medical imaging department of the local hospital would be very worthwhile.

◆ *Further activities*

- ◆ Binaural hearing is a good subject for a formal investigation, with opportunities for hypothesising, consideration of variables, and measuring and recording results.
- ◆ Pupils could cup their hands around their ears and find out how much the increase in collector size improves the ear's sensitivity and directionality.
- ◆ Pupils could use a sound level meter (see the *Equipment notes*, page 95) to record different sounds, or to investigate how loudness changes with distance from a source (intensity falls according to the inverse square law but see the point about loudness in the *Enhancement ideas* opposite). An alternative way to compare the loudness of sounds is to measure the distance from the source at which it just becomes inaudible.
- ◆ To assess whether pupils have got a functional understanding of how the ear works, ask them to design or assemble on paper a 'bionic' ear from satellite dish, drum, set of levers and microphone, explaining how each device relates to a part of the ear.

◆ *Enhancement ideas*

- ◆ Any teaching about hearing and its impairment must of course be done with sensitivity to those whose hearing is impaired. One approach is to make use of the experience of such pupils to find out how they manage to overcome their handicaps, and how others can help. For example, pupils could learn some British Sign Language to communicate with profoundly deaf people. Alternatively, join up with a PSE lesson and invite a guest speaker to talk about deafness, hearing dogs and so on.

- ◆ Teenagers do not, in general, take kindly to warnings from parents and teachers that their lifestyles or aspirations could be hazardous. In the cause of protecting their hearing, pupils might be more impressed by the testimony of pop musicians. One musician whose career was ruined by loudness-induced deafness has set up an organisation HEAR (Hearing Education Awareness for Rockers) to warn of the dangers. Their web site at **www.hearnet.com** is worth visiting.

- ◆ A study of animal ears will reveal some interesting variations from humans, apart from frequency range. The barn owl has a height indicator for locating sound sources (such as dinner!) – the feathers around the ears are funnelled upwards and downwards. Pupils could try this using paper ear trumpets on their own ears.

- ◆ Loudness of sound is related to the amplitude of the source's vibration. This means it is related to energy and intensity (energy per unit area). When sound levels are measured, however, a comparative scale is used. Each sound's level is compared to a threshold level at which a sound can just be heard. This is about 10^{-12} W/m^2, or 10^{-17} W on the eardrum – showing the ear's extraordinary sensitivity. This sensitivity to sound depends on changes in the intensity, that is to say we notice as equal changes every doubling of intensity. The sound level scale reflects this. The bel scale, better known in tenths as the *decibel scale*, is logarithmic with respect to intensity. Doubling the intensity of the sound will give a sound level increase of 3 dB; a 10 dB increase represents an intensity increase of 10 times. This scale tends to mask the potential hazard of loud sounds. When the sound level doubles, the safe exposure time is halved, according to the Health and Safety Executive (Table 2.1). Table 2.2 shows some typical sound levels.

Table 2.1 *Safe exposure times.*

Loudness in decibels (dB)	Maximum safe exposure time	Comment
90	8 hours	conversation difficult, need to shout
93	4 hours	
96	2 hours	
99	1 hour	conversation possible only at top of voice
102	30 minutes	
105	15 minutes	conversation impossible

Table 2.2 *Sound levels (at 1 m from source).*

Sound	Loudness in decibels (dB)
ticking watch	20
conversation	55
background music	65
alarm clock	80
Wembley concert	up to 105 at 50 m
club sound system	up to 120

♦ Medical uses of ultrasound have increased rapidly over recent years. For imaging it has the advantages over other methods of being inexpensive, non-invasive and without side effects. This means it is widely used in obstetrics, and for investigating such things as the presence of gallstones. The ultrasound probe is a transmitter and receiver. The method involves producing reflection from organs in the body. Imaging of soft tissues is particularly good compared to radiological methods. Recent computing developments have led to improved scanning, storage and display techniques. Note that many pupils have seen such images on television; they tend to assume that the waves used are electromagnetic rather than sound, and they may believe that a hole must be cut to let in the waves.

2.3 Nature and transmission of light

◆ *Previous knowledge and experience*

Children in primary school will have had experience of distinguishing between primary and secondary sources of light, and been encouraged to realise that it is light, rather than darkness, that is an entity. They will have investigated shadows, especially in the context of a Sun clock, and perhaps looked at relationships between position and size. They will have looked at the interaction of materials with light, and perhaps be able to use the terms *transparent, translucent* and *opaque* correctly. They will have had some experiences of using mirrors to reflect light and of studying images in mirrors – through their own toiletry requirements and with fun curved mirrors! They may have studied the eye as a receiver of light. The video *Making Sense of Science: Light* shows examples of good primary practice (SPE, 1996).

At the end of the primary phase many children's conceptions of light are far from those that a physicist would accept. They consider light in terms of its source, its effects or as a state. This permits a descriptive account of the situations and classroom experiences. What is often lacking is the idea of light as an entity located and *travelling* in space. Light may be thought of by pupils as a kind of 'bath' of energy in which we are immersed, with no sense that it is moving from place to place. It may not be conserved, it is only noticed when it is intense, and it can be intensified by a magnifying glass, for example. Mirrors may be thought to 'hold' light to make an image, and matt surfaces may not be considered as light reflectors. Some pupils will use the words *reflection* and *shadow* interchangeably. Even if light is said to travel – say, as a lighthouse beam – there is no acknowledgement of propagation time, which is unsurprising given its high speed. Children at the secondary phase need to develop the concept of light travelling, as a beam or a ray, before they can use this in optics explanations. For many pupils at the age of 13 or 14 this concept will still be developing, and they may have difficulty comprehending an idea such as a virtual image. Few pupils will have been taught refraction and the action of lenses at primary school, but they can give descriptions of the effects such as 'water makes the pencil look broken'.

♦ *A teaching sequence*

Given the above, it is a good idea to give pupils lots of experience of light as an entity. This means that if at all possible the class should have blackout facilities. Torch or overhead projector beams or, better still, a laser beam (see the *Equipment notes*, page 95), can be sprinkled with chalk dust to show their shape and extent. The representation of these with ray diagrams should be introduced carefully if pupils are to assimilate this concept. The contexts of media, advertising and entertainment's use of lighting effects can be motivating and instructive. The use of software and audio-visual aids will also make valuable contributions to this topic.

Finding light

Establish first that pupils can distinguish luminous from non-luminous objects, using a range of examples, in the classroom. Electric lamps (filament, LED and fluorescent), flames and the Sun are obvious sources. Pupils may be less sure of the status of the Moon, fluorescent substances, mirrors and shiny materials, as they may confuse reflection with generation of light. Pupils could complete a simple grouping chart and discuss those they are unsure of.

Out of this discussion two important points may arise: that light cannot be seen in transit, and that it appears to take no time to travel. This is the opportunity to show beams of light shining on chalk particles or similar. The straight-line travel of light can also be noticed in this situation. Reference could be made to 'sunbeams' which are only visible when there is mist or dust to show them up. This effect is simulated at pop concerts and discos; most pupils will have seen the effect on TV programmes such as *Top of the Pops*, which may have supported the misconception that you can actually see light itself.

The speed of light is nearly 300 000 kilometres per second (3×10^8 m/s), so it appears to travel instantaneously over terrestrial distances. Most international telephone links now use light travelling along optical fibres; phone conversations around the world show no noticeable delay. In the vastness of the Universe it is different. The light reflected from the Moon takes 1.5 seconds to reach us, light from the Sun takes about 8.5 minutes. From the nearest star the light has travelled for over 4 years, and from the farthest point of the Universe

15 billion years – the age of the Universe itself. This means that looking into space is looking back in time. When we look at a star in the night sky, we are seeing it as it was years – perhaps millions of years – ago. The idea may not help pupils get a feel for the speed of light, but they will be stimulated by the idea of time travel.

If pupils have studied sound recently, comparison with the speed of sound can be made, with examples such as thunder and lightning, and the use of a starting pistol for races. To calculate the distance of a thunderstorm, time the gap between lightning flash and thunderclap and divide by 3 (for km) or by 5 (for miles).

Materials and light

Pupils can extend work that they have probably begun in primary school, by classifying objects as opaque, transparent or translucent. A range of materials can be supplied for placing in front of a torch or ray box, and monitored with a light sensor. The activity can be used to direct pupils' attention to the nature of light; when it interacts with materials it can be reflected, absorbed or transmitted. Many materials cause a combination of these. For example, ordinary window glass reflects as well as transmits light, as pupils will realise if they look out of a lighted room into the night. This situation can be set up with a safety screen in a darkened laboratory. (See also 'Pepper's Ghost' under *Further activities*, page 71).

It is the interaction of white light with materials that causes colour. Coloured opaque objects reflect the colours we see and absorb the rest of the spectrum. Coloured transparent and translucent materials transmit the visible colour and absorb the rest. Some substances produce unusual colour effects. A dilute solution of milk scatters blue light more than red. Dyes such as fluorescein or eosin reflect some colours, and absorb and then re-emit others (this is fluorescence).

Pupils need to realise that not only mirrors or smooth surfaces reflect light. These give *regular* reflection and can produce images. Almost all non-luminous objects are visible by reflected light, but this is *diffuse* reflection. The model of bouncing a ball off a wall can be used to explain the difference. If the wall is smooth we can predict the way it will rebound. If it is rough the ball will bounce unpredictably in many different directions. Matt surfaces are microscopically rough, so light reflects off in all directions and no image is formed.

When light travels through a medium such as air or glass, it will meet dust particles and other imperfections such as variations in density. These scatter the light. Blue light is scattered more than red which is why most of the sky is blue; sunsets and sunrises are often red because we are looking at the Sun's light which has not been scattered.

Shadows

Shadow activities bring together the ideas of opaque materials interacting with light, and the straight-line travel of light. Pupils will have carried out some general investigation of shadows in primary school; this can be built on in several ways. The geometry of shadows can be explored, perhaps in the context of making model puppets and a theatre. (This could be an activity for a science club.) Alternatively, they could look at multiple shadows produced with several sources. This is easier to sort out if the sources are different colours.

A development of this is to explain why an object has regions of total shadow (umbra) and partial shadow (penumbra) when illuminated by an extended light source. An example of this is in an eclipse of the Sun by the Moon (Figure 2.10). Pupils should appreciate that rays are coming from the Sun in all directions, but we need only draw four to locate the umbra and penumbra.

Figure 2.10
Explaining an eclipse of the Sun.

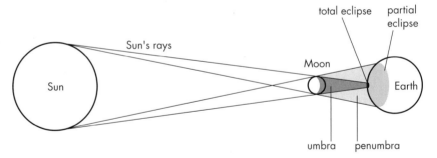

How light travels

The traditional activity for showing rectilinear propagation of light involves lining up two cards, each with a hole in, and shining a light through onto a screen. This is best done as a demonstration, threading string through the holes from source to screen to show the straight path (Figure 2.11). This then becomes a model of a light ray, an important concept in the explanation of optical effects. In optics, a ray is an idealised path of light which can be represented as a line, so the geometry of optical behaviour can be explored. In reality the path of light always spreads out as it travels from a source; this is a light beam. A laser is a good visual aid for this as the beam does not spread very much. Pupils could try to look down rubber tubing, and experience the need for it to be straight.

Figure 2.11
Light travels in straight lines (rectilinear propagation).

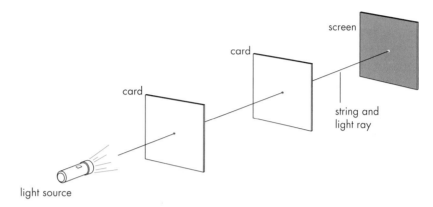

This is a good opportunity to explore pupils' ideas of how we see. Asking 'Why can we not see around corners?' suggests an active role for the eye which is a common misconception in younger children (Figure 2.12). Get pupils to draw their explanation, using arrows to show how light travels from an object which can be seen and one which is obscured. Some are likely to draw light rays coming from the eyes; this idea is not helped by comic-book ideas of 'X-ray eyes' and the like. Develop this discussion into an exploration of visibility, for example of road signs or cyclists' clothing (see *Further activities*, page 71). This should help to elicit whether pupils can accept that most things are visible by diffuse reflection, and that visibility will depend on the intensity (strength or brightness) of the reflected light when it meets the eye.

Figure 2.12
Children's drawings of how we see. (Source: Primary SPACE Research Report: Light by Osborne et al, 1990. University of Liverpool Press.)

Mirrors

Mirror reflections can be introduced using a torch and a large mirror in a darkened room. Can pupils predict where the beam will be reflected, as the angle of the beam to the mirror is changed? Most will have an intuitive feel for this, which can be substantiated with the use of a ray box and mirror on the bench (Figure 2.13). Many pupils will need the support of a worksheet to collect and interpret the data leading to the law of reflection.

Figure 2.13
Using a ray box to investigate reflection.

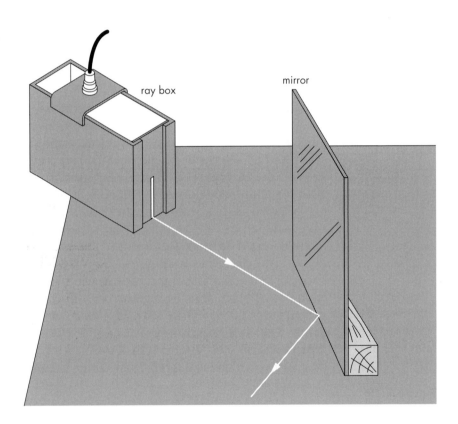

Pupils should be able to draw a ray representation of the law of reflection, using the normal to the plane surface (Figure 2.14a). Faster and more able pupils can try to apply this law to the reflections they observe with curved mirrors (Figure 2.14b). Pupils will have to learn that we consider angles *relative to the normal*, the line drawn at right angles to the surface; it is not obvious why we do this when considering a plane surface, but for a curved or irregular surface it makes more sense.

Figure 2.14
*The law of
reflection.*
a *Reflection at a
plane mirror.*
b *Reflection at a
curved mirror.*

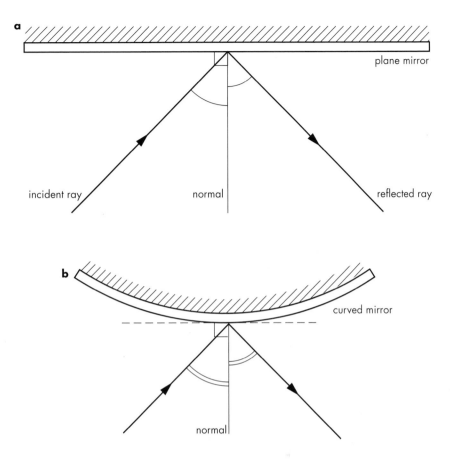

Pupils enjoy exploring the images in a pair of mirrors, set at different angles to each other (Figure 2.15, overleaf). With guidance they should be able to deduce a pattern relating the angle between the mirrors (A) and the number of images (N). This is:

$$A \times (N + 1) = 360$$

For example when the angle is 90° there are three images; one in each mirror and a third one formed by the reflection of each of these in the other mirror. Pupils should look at their own face in the two mirrors at right angles to each other. This is the face everyone else sees. It is not inverted left to right, like the usual image of themselves. Can they explain how it is different? Can they comb their hair while looking in such a mirror?

This is also an opportunity to look at symmetry, for example by placing a mirror vertically across the middle of the letters of the alphabet, words like CARBON DIOXIDE, or other figures.

The image formed by a plane mirror is as far behind the mirror as the object is in front. This is odd since light cannot get through the mirror. It is an example of a rather abstract concept – the *virtual image*. Light *appears* to be coming from a point behind the mirror, when in fact it has bounced off the front of the mirror. Pupils need to learn to distinguish between our brain's interpretation of what we see (an image behind the mirror) and what is actually happening, in terms of light rays.

You may also encounter the age-old question of why a mirror reverses left–right but not top–bottom. In fact, it would be more correct to say that a mirror reverses back and front. Try writing a word on an OHP transparency. Hold it up to the mirror and look through it at its reflection. Both words look the same. The problem is that we normally picture ourselves going round behind the mirror to occupy the space where our reflection is. As we do this, we turn round through 180°, thus reversing left and right.

◆ *Further activities*

◆ Ask pupils to compile a record of lights being used for communication, say on their journey to and from school. These could include traffic indicators of various kinds, travel information screens, advertising and warning signs. How is the light generated, how is the message transmitted, and what kind of control is needed?

◆ For homework pupils could research the history of telecommunications, or the advantages and disadvantages of using light to communicate over long distances.

◆ Investigate the factors affecting the visibility of objects, for example for road safety. Light intensity could be compared using a light meter or sensor, or measuring the distance to invisibility. Different colours and surfaces can be tried, especially the reflective material used in cyclists' clothing, etc.

◆ Kaleidoscopes use the multiple symmetry of angled mirrors to produce attractive patterns. Pupils could investigate how many mirrors, at what angles, produce the best effects.

◆ Check whether pupils have an understanding of the law of reflection and simple ray diagrams by asking them to design and build periscopes. To aid pupils, a cut-out design of a periscope could be drawn on a worksheet.

◆ 'Pepper's Ghost' is an optical illusion using the fact that glass both reflects and transmits light. For example, a virtual image of a lighted candle can appear to be submerged in water (Figure 2.16).

Figure 2.16
The Pepper's Ghost illusion.

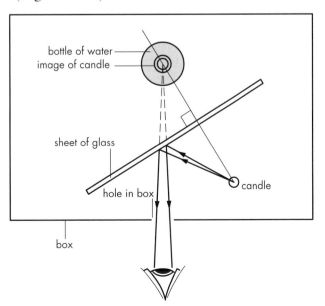

◆ *Enhancement ideas*

- ◆ The filament lamp is an inexpensive, everyday object which we take for granted. Its development was something of a scientific epic, however. Two famous entrepreneurs were in competition – Joseph Swan in Britain and Thomas Edison in the United States. The problem was to find a suitable material for the filament. It had to give out as much light as possible, by being extremely hot, without burning or oxidising. Lewis Latimer, an African-American scientist, finally found the key – a fine carbon filament in an evacuated bulb. Even in modern tungsten filament lamps only 3% of the input energy is transformed to light, the rest is radiated as heat. Later developments in lamp technology have increased this to more than 20% in fluorescent discharge tubes.
- ◆ The producers of science fiction films can contribute to the misconception that light is visible. In the *Star Wars* films weapons emit light, and to make the effects dramatically visible the light beams had to be painted in, and toy weapons were provided with fluorescent tubes to show the light beams!
- ◆ An attractive example of the use of shadow puppets is the traditional Indonesian technique which uses ornate sheet metal figures to produce shadows on a translucent screen. Stories of gods and people are accompanied by a gamelan orchestra and, by moving the figures away from the screen, the gods became enormous compared to the human figures!
- ◆ Laser light is produced by exciting atoms electrically so that they emit light energy in a very *coherent* way – the light waves from every atom are 'in step'. Although a laser is not necessarily of high power, its beam of light does not spread, so that it maintains a high intensity. It is this that makes it potentially dangerous. The brain's normal reflex defence to a bright light is not always activated quickly enough and small patches of the retina can be damaged.

  The coherent nature of laser light makes it useful for telecommunications. The light is modulated by an electrical signal, e.g. from a telephone mouthpiece, and the varying light beam then travels along optical fibre with very little loss of energy. The digital light signals are easily cleaned up to remove noise, before being transformed into sound again at the receiving end (Figure 2.9, page 59).

2.4 Vision and the electromagnetic spectrum

Previous knowledge and experience

The idea that we see with our eyes giving out something, rather than receiving light, is still prevalent amongst some secondary pupils (Figure 2.12, page 67).

Analysing vision and its defects can provide a good context for studying the action of lenses. Pupils need to be familiar with the ray convention for showing how light travels. This could be introduced by demonstrating with a laser or slide projector to give intense narrow beams, which are seen when they scatter light into the eye off dust particles (see Section 2.3).

Pupils will know that white light can produce a spectrum; this can be developed in relation to refraction through transparent media. Pupils will also need to explore combinations of coloured lights, pigments and filters.

A teaching sequence

The eye and lenses

Pupils enjoy testing their own eyesight in simple ways. Some tests to try:

- finding the blind spot;
- watching the dilation of a partner's pupils when they open their eyes;
- testing binocular vision by trying to pick up small items with one eye closed.

Ask pupils to focus successively on near and far objects and explain that what they experience is the sensation of changing the shape of the eye's lens. Show a model eye and locate the lens. A model eye can be made from a large (e.g. 5 litre) round-bottomed flask, filled with a dilute solution of fluorescein, to make the path of the light beam visible (Figure 2.17, overleaf). Three lenses are attached with Plasticine to the surface of the flask, for normal vision, short and long sight. With a 5 litre flask these will need to have powers of $+8D$, $+11D$ and $+5.5D$, respectively. (D stands for dioptre.) When a strong, narrow beam of light is shone through each lens in turn the beam is seen to come to a focus on the back surface, in front and behind, respectively.

Figure 2.17
A model eye.

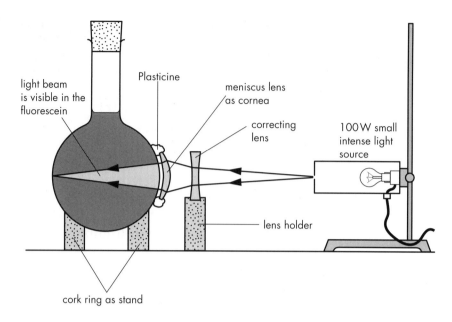

light beam
is visible in the
fluorescein

Plasticine

meniscus lens
as cornea

correcting
lens

100W small
intense light
source

lens holder

cork ring as stand

This shows how the strength of the eye lens needs to increase for nearer objects, to 'focus', that is, produce a sharp image on the back of the eye (Figure 2.18). In a real eye, the lens when relaxed is thin and focused at infinity. The ciliary muscles contract and fatten up, squashing the lens into a fatter shape, to see near objects clearly. (Note that, in a real eye, much of the focusing happens at the surface of the cornea, where the light first enters the eye. The lens provides a fine adjustment to this.)

Figure 2.18
Light is focused on the retina.

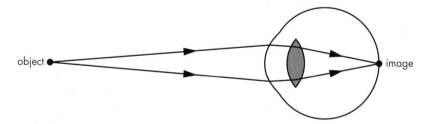

object

image

The model eye can also show the effect of spectacle lenses in producing focused images. In long sight the corneal curvature is not strong enough to produce a focused image on the retina, so an additional *converging* lens is used. In short sight, the converging effect is too strong for the eyeball, so this is corrected with a *diverging* lens (Figure 2.17). Pupils should handle the lenses to distinguish converging lenses from diverging. It is best to avoid the terms convex (curved out) and concave (curved in) as many lenses, including those used in spectacles and as contact lenses, have one surface of each type. It is better to concentrate on the function. Converging lenses are thicker in the centre; they make parallel rays converge. Diverging lenses are thinner in the middle; they make parallel rays diverge.

Convergence and divergence of rays can be shown using a ray box and *cylindrically* curved lenses (in contrast to the normal ones which have *spherically* curved faces). It is unlikely that pupils will be able to see where the light travels through the lens, but they should draw the directions of the rays up to and beyond the lenses. The strength of a lens, its power, is related to the curvature of the lens surface. It is inversely proportional to the focal length. This is the distance from the lens that a parallel beam of light comes to a focus (Figure 2.19). Pupils should find the focus of different strength lenses with a ray box and then with spherical lenses and a distant light source. They should represent their observations with a simple ray diagram like those in Figure 2.19.

Figure 2.19
Defining the focal length of a lens.
a *A converging lens.*
b *A diverging lens.*

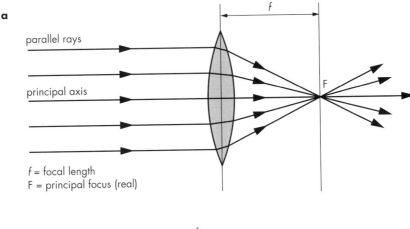

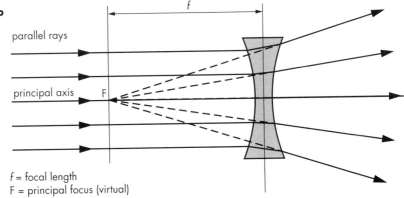

Refraction

Pupils should explore how light travels through a parallel-sided glass block, to understand the nature of refraction. They should change the angles of incidence of the ray box beams on this block and also a semicircular block to establish the following (Figure 2.20):

- the change of direction occurs only at an interface;
- light that travels perpendicular to an interface is not refracted;
- light bends towards the normal (inwards) when it travels from a less dense to a denser medium, and vice versa;
- when light travels to an interface with a less dense medium it can be reflected (total internal reflection).

Figure 2.20
Paths of light rays through a glass block.

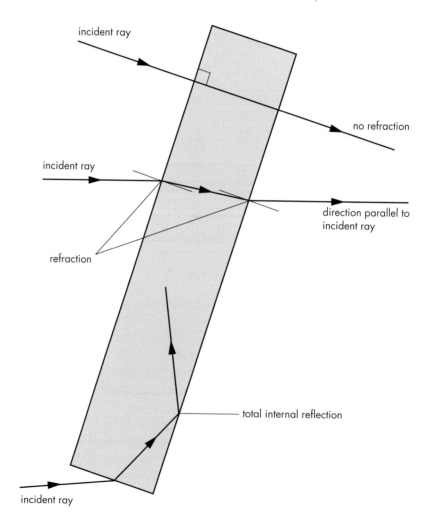

The camera

The eye is often compared to the camera. Pupils can compare the methods of focusing, aperture and shutter control.

The *pinhole camera* provides a good opportunity to reinforce the function of a lens in gathering light to a focus. The camera consists of a box with a front which can be pierced and a rear made of a translucent screen. With a single pinhole, a sharp image of a bright object is produced. (Use carbon filament lamps for the best results.) Pupils can draw a ray diagram to show why this image is inverted (Figure 2.21). Further pinholes produce multiple images; a large hole causes blurring. If a lens of the correct focal length is placed over the large hole the image becomes sharp again, but brighter than the pinhole image.

Figure 2.21

Forming an image with a pinhole camera.

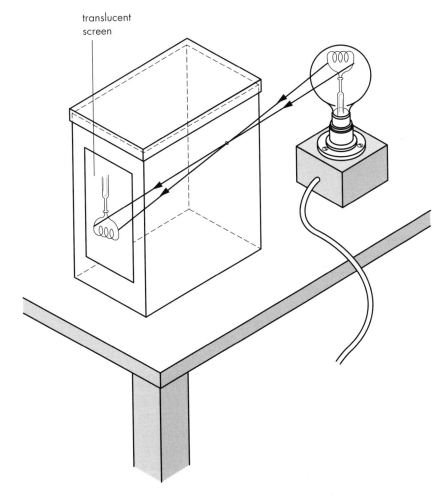

A simple telescope

A simple telescope can be set up with a pair of lenses, suitably spaced along a metre rule (Figure 2.22). A carbon filament lamp is ideal as the object, or a bright outdoor scene. It is best if the rule can be supported at eye level. The lenses can be held in place by Blu-tack if lens holders are not available. A relatively weak converging lens acts as objective lens (focal length about 50 cm, power +2D). This forms a real but inverted image of the distant object, 50 centimetres from the lens. This can be 'caught' on a translucent, tracing paper screen. The eyepiece lens is stronger with a focal length of about 10 cm (+10D), and forms a virtual (upright) image of the first image. The two lenses are thus 60 cm apart. The translucent screen is removed and the image of the lamp is seen directly. The magnification is 50/10 = 5. Pupils may need help to see this, and the teacher should adjust the lenses, to suit. When they do see it there is a good sense of achievement.

Figure 2.22
A simple telescope; remove the screen to see the final image.

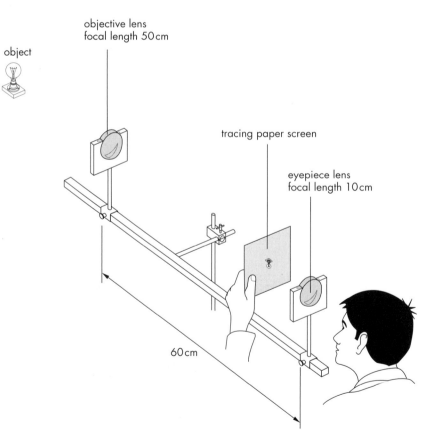

objective lens
focal length 50cm

object

tracing paper screen

eyepiece lens
focal length 10cm

60cm

Colour from light

Demonstrate the action of a prism in producing a spectrum from white light. This is best done in a darkened room using a strong source such as a slide projector. Hand out prisms and let pupils explore the images they see through them. In some positions they will see spectral fringes (window bars are good for this). They will also see reflected images without any colour fringes. It is too complex to try to explain the various images and spectra seen. However, you should set up a prism in front of a ray box to demonstrate the appearance of a conventional spectrum (Figure 2.23). A second prism can be used to repeat Newton's renowned experiment to recombine the spectrum to give white light. Ask pupils to record the sequence of colours, from the least refracted (red) to the most (violet). Can they really see the seven colours that Newton originally recorded? Most people cannot separate the indigo, and it is thought Newton included this because of his belief in the mystical significance of the number seven.

At this point you could also demonstrate Newton's colour wheel, a spinning wheel marked with segments in the spectral colours. They combine to give a greyish colour, rather than white, because of their impure quality. Pupils could make their own wheel, and spin it on a string passing through two holes on either side of the centre.

Figure 2.23
Using a prism to produce a spectrum.

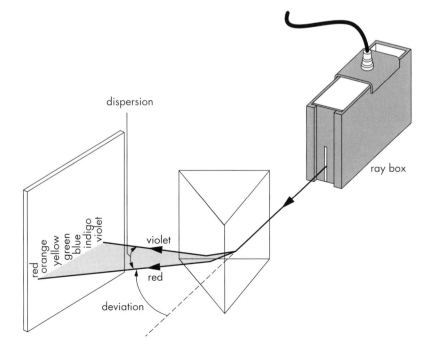

Pupils can insert colour filters into their ray boxes and observe how these remove some of the spectrum of the prism, e.g. red only allows red to pass, cyan allows blue and green. They can then investigate the mixing of coloured lights produced by the filters, either with ray boxes or torches. They will need to record their findings carefully if any pattern is to emerge. Primary colour filters are pure red, green and blue. With these any other spectral colour can be produced by colour addition. The secondary colours are cyan (blue and green), magenta (red and blue) and yellow (red and green). However, most filters allow through a range of colours, so results are often unconvincing. Filter suppliers often provide details of the transmission characteristics, which pupils may find interesting.

TV screens consist of dots of phosphors of the primary colours, which light up when the electron beam in the TV tube hits them. They can be seen with a magnifier.

When considering the apparent colour of objects, reference could be made to variations seen in daylight and artificial light, e.g. the colour of clothes when buying in brightly lit shops, and the appearance of cars under street lights. It is worth providing matt silver and white objects which will reflect all light but diffusely and not produce images (as this requires a polished surface). Discussion about visibility of colours in the context of the lighting levels can lead into a consideration of colour vision (see *Enhancement ideas*, page 83).

Beyond the visible spectrum

A light dependent resistor in a circuit can be used to show that the spectrum extends beyond the visible. Connect up the circuit shown in Figure 2.24, or use a manufacturer's detector, to explore the readings through a spectrum produced from a prism and slide projector. Show that energy is radiated beyond red (infra-red) and beyond violet (ultraviolet). Discuss with pupils their experience of each.

Figure 2.24
Circuit for detecting infra-red and ultraviolet radiation.

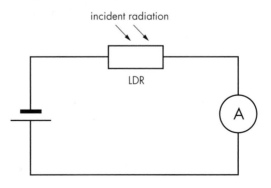

incident radiation

LDR

A

Infra-red

Provide a hot object (e.g. radiator or non-glowing hotplate) and talk about what is felt. This infra-red (IR) is what is popularly called 'heat rays'. It is better to use the term infra-red radiation and to say that it represents a transfer of energy, as does the whole of the electromagnetic spectrum. When matter absorbs this energy its temperature rises. The energy is then in the form of increased movement of the particles of matter (internal energy). IR beams are used in remote control systems such as TV channel 'zappers' which pupils will have used. The rays do not feel warm because they are very weak. There is a circuit in the TV which is similar to that in Figure 2.24 to operate switches.

Ultraviolet

Use an ultraviolet lamp to illuminate a range of objects including fluorescent rocks and white cotton. The lamp is likely to be visible through emitting violet light; true ultraviolet (UV) is invisible. Its photons (see Section 2.5) are more energetic than those of visible light.

> **!** *Use a screen around the UV lamp, so that pupils are not exposed to its direct rays. All UV may cause skin cancer and short wavelengths may cause eye damage.*

Fluorescence occurs when UV is absorbed and some energy is re-emitted in the visible range. Most detergents contain fluorescers, called 'optical brighteners', so that when washed clothes are illuminated by white light some UV radiation is absorbed and then re-emitted as visible light, causing the 'whiter-than-white' effect, beloved of advertisers. Pupils may have experienced UV lighting at discos and theatre performances and will enjoy fluorescent effects if blackout can be provided. Some false teeth appear black under UV!

The whole spectrum

Show pupils a chart of the whole electromagnetic spectrum, and discuss the similarities and differences in the nature and uses of the various radiations. A suitable chart is published by the Pictorial Charts Educational Trust (PCET) – see *Other resources*, page 97. The main points of similarity are that all radiations travel in all directions, in straight lines, at the same speed in space (c is about 3×10^8 m/s). The main differences are as a result of their photon energies. X and gamma radiation are most energetic so they can penetrate matter (including human tissue) depositing their energy as they penetrate. This can cause chemical changes with biological consequences, such as the initiation of cancer or genetic damage which may be passed on to future generations. Radio waves have least energy, are not readily absorbed and therefore have little effect on matter. The relationships between speed, energy, wavelength and frequency are covered in Section 2.5.

When discussing the electromagnetic spectrum, note that frequency ranges are only approximate; the divisions of the spectrum are not well-defined.

Pupils could research the uses of different electromagnetic radiations. Each group could look at one from radio, microwave, infra-red, ultraviolet, X-rays and gamma rays, and report on how they are generated, detected and used, and on any hazards in use. Make your own large wallchart and ask all pupils to contribute.

◆ *Further activities*

- ◆ Pupils could bring in cameras or research camera catalogues to compare specifications and report on them, perhaps in the form of a *'Which?'* consumer guide. Which features are common to all? Which features offer valuable improvements in picture-taking, and why?
- ◆ The video camera is a better model for the eye, in that it operates continuously and converts the light of images into electrical signals for storage. Ask pupils to write an account of the camcorder, drawing attention to the similarities to and differences from the eye.
- ◆ Pupils could investigate the size and nature (inverted or upright) of images produced by different strengths and types of lens. They could examine how additional magnification can be obtained by combinations of lenses in the microscope and telescope. A simple telescope can be set up by pupils (see Figure 2.22, page 78).
- ◆ Spectacle lenses are designed to do several things. Pupils could investigate the images they produce, or find their focal lengths. Spectacle prescriptions are quoted in dioptres; power in dioptres (D) = 1/focal length in metres. Converging lenses (for long sight) have +D values, diverging lenses (for short sight) are −D. Usually lenses are convex on the front surface, even if they are diverging. Older people often require focusing assistance at both short and long distances, so 'bifocal' or variable focus lenses are used. Spectacle lenses can contain a substance that darkens in strong light ('Reactolite' is one brand), so they protect

against glare. Polaroid is another way of protecting. This makes use of the wave nature of light (see Section 2.5) to absorb much of the scattered light from objects. Pupils could research the design of spectacles, including, for example, suitable materials and shapes for use in sports.

◆ The effect of coloured light on coloured objects is best investigated in boxes provided with holes for illumination and viewing. This helps to avoid light reflected from other objects. Even so, the results often conflict with simple theory because of the contribution of the brain in vision – we know tomatoes are red even if there is no red light reflecting from them!

◆ As an extension to the work on colour mixing, some pupils could light a model stage with low voltage lighting and filters. They could experiment with combinations of colours to give different dramatic effects – blue for cold/night, red for anger/firelight, etc.

◆ *Enhancement ideas*

◆ Total internal reflection is made use of in fibre optics, the modern alternative to electrical cables for telephone, cable TV, ISDN lines, etc. They consist of very thin, pure silica fibres, coated to give good internal reflection, and bundled together to increase message-carrying capacity. The advantages of optical fibres over conventional cables are the very much higher information-carrying capacity, reduced interference, and reduced size and cost. In addition the signals are carried in digital form and can travel further without requiring amplification and noise reduction, resulting in further cost-savings. Fibre optics also feature in endoscopes, the viewing devices used for internal examination and 'keyhole surgery' of the human body. Pupils may be familiar with decorative lamps which use optical fibres.

◆ Ask pupils where spectra of light occur naturally and discuss how the formation of a rainbow might be similar to the action of a prism. (The water droplets act as prisms; light refracts in them and then reflects towards the viewer, so the rainbow is seen when the Sun is behind you.) Note that the colours seen in soap films and oil films on water occur for a different reason. These are caused by the interference of light (see Section 2.5).

◆ Pupils are often fascinated by colour blindness charts ('Ishihara tests'), although this will need some sensitivity as one in 12 males has some impairment of colour vision, usually red/green confusion. (In France this impairment is called *daltonisme*, after John Dalton, a sufferer who also invented the modern idea of the atom.) Colour vision is provided by the cone cells in the retina. These come in three types which respond to the three primary colours of light. They need a higher level of illumination than the rod cells, which distinguish black and white. Thus we can only see in shades of grey when there is very little light.

◆ The effects of IR and UV radiations on the human body differ because of the energy of their photons. IR raises the body temperature, usually locally, and can cause a burn, possibly a painful blister. UV causes 'sunburn'. It penetrates the external surface of the skin so that the underlying layer can be damaged or even killed, and it flakes off, perhaps painfully. UV can also cause chemical changes in the skin. A common result in some pale-skinned people is the release of melanin, causing 'suntan'. Unfortunately, UV can also cause skin cancer, which usually shows as a growing dark spot or mole, called a melanoma. Loss of ozone in the upper atmosphere has caused an increase in the UV exposure from sunlight, so it is even more advisable to protect our skin from this with sunscreen oils. UV is quite easily absorbed by any substance and is absorbed by glass or water.

◆ Infra-red radiation was discovered around 1800 by the renowned astronomy team of Caroline Herschel and her brother William. It is now used for astronomical observation as it provides different information from visible radiation. Likewise radio telescopes and X-ray receivers help to build up our picture of the Universe from their imaging.

◆ The discovery of X-rays by Wilhelm Roentgen in Vienna in 1895 is a fascinating story, both in the development of scientific understanding, and in its immediate medical and social impact. Within days of the news reaching America, X-rays were being used to locate a bullet in a patient's leg. Meanwhile there were letters to *The Times* protesting about the impudence of scientists who could now see through a person's clothes!

2.5 Waves

◆ *Previous knowledge and experience*

Pupils will probably have been introduced to ripples on water as a wave model for light, and to a slinky spring for modelling a sound wave, during the primary or early secondary years. These will be reviewed and developed in this section. There is a need to distinguish between the regularity of wave motion and the idea of a wave pulse. Both sound and electromagnetic waves are usually continuous, whereas seismic waves are usually pulses. A pupil's idea of a wave is likely to be of seaside breakers for surfing. These have more in common with pulses and they have some rather unusual features, which make them a poor model.

Pupils may associate the movement of a wave with energy. One of the goals of this section is to help them realise that though energy is transferred by a wave, the medium of transfer is not transferred. (Waves travel across the sea, but the water remains where it was once the wave has passed.)

◆ *A teaching sequence*

This section brings together all the theoretical aspects of wave study which may be studied in the secondary curriculum. This theory – and modelling – is applied to a range of situations within pupils' experience. Instead of being taught as a unit on waves, parts could be introduced at an appropriate stage when teaching sound, or light, or the electromagnetic spectrum (see *Choosing a route* at the start of the chapter).

Wave models

A source of sound is a good place to start this topic. It should be clear to pupils that a sound is the result of a continuous vibration. A tuning fork could be shown as a classic example of this. A slinky spring is a model in which small volumes of air are represented by the loops of the spring. As a tuning fork moves outwards, in the model the spring is pushed together (Figure 2.25, overleaf). This compression is then transmitted along the spring as a pulse. It is followed by a steady sequence of pulses, one for each oscillation of the fork. Each loop of the spring simply moves back and forth, being part of a *compression*, then its opposite, a *rarefaction*. Pulse movement can be seen easily with a slinky spring, but the movement of an individual loop is harder to follow. This is because it moves in the same direction as the wave pulses. This is called a longitudinal wave, in contrast to transverse waves, where the movement of the medium is perpendicular to the wave direction. Challenge pupils to make this second type of wave with a slinky. Electromagnetic waves and water ripples are transverse.

Figure 2.25
Sound waves compared to a mechanical longitudinal wave on a spring.

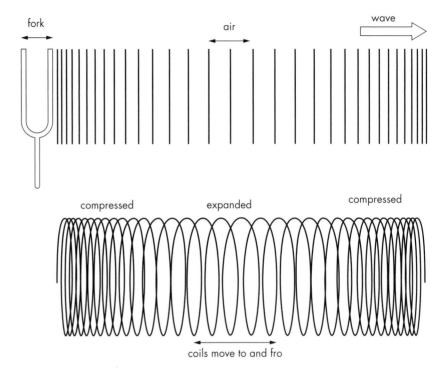

Dip a finger in a tank of water to generate ripples. Direct attention to the way that the surface of the water moves up and down, while the ripples move outwards. Pupils can see this more clearly if cork, polystyrene or paper 'floats' are scattered over the surface. Anglers in the class will be very familiar with this, having had to distinguish such movement from the tugging of a catch on the line!

A heavy rope or length of rubber tubing is a good model to show transverse pulses. One end of it is moved up and down, simulating the way ripples in water are generated. For each up-and-down movement (oscillation), a pulse emerges and travels along the medium, i.e. at right angles to the oscillation. The easiest way to get a good pulse is for two pupils to hold each end in slight tension and for one to give a rapid flick to the end (Figure 2.26).

Figure 2.26
Making a transverse wave pulse.

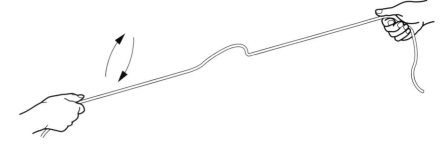

If pupils keep hold of the rope at both ends they can observe how the pulse is reflected back. Producing a stream of pulses will develop a stationary pattern, called a standing wave. This shows the key features of a wave. The upward bulge is called the wave crest, and the corresponding downward one is a trough. The distance from a point on one crest to the next similar point is the wavelength, λ (Figure 2.27). A larger flick increases the height or amplitude a of the pulse – this is associated with more energy. The rate at which the pulses are generated is the frequency, f. If the frequency is increased then the wavelength can be seen to decrease. You will need to define these quantities carefully for the class. Points to stress:

- amplitude, a, is measured from the centre line, not from crest to trough;
- wavelength, λ (Greek letter lambda), is measured between any two corresponding points, e.g. trough to trough;
- frequency, f, is not to be confused with wave speed.

Figure 2.27
Defining amplitude and wavelength.

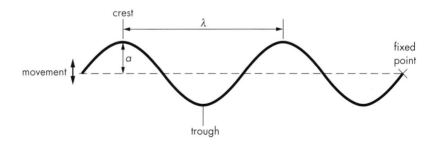

At this point it would be helpful to look at the oscilloscope traces produced by sound waves (see Section 2.1). The oscilloscope is plotting a graph of voltage against time. The voltage represents the displacement of the air by the sound vibration. So the y-axis represents the displacement of the sound wave. If the timebase is set so that a fixed trace is obtained, then the frequency can be calculated.

In Figure 2.28, the frequency *f* is 2 oscillations per second, i.e. 2 Hz. Although Figures 2.27 and 2.28 have the same shape, two significant differences need to be pointed out to pupils. Figure 2.27 is a position graph for a transverse wave. It shows the horizontal versus the vertical displacement at a point in time – a slice through the wave. Figure 2.28 is more complex. It is a graph of how the displacement of a particular point in the medium *changes with time*. This is how a longitudinal sound wave produces a graph showing a transverse variation.

Figure 2.28
Working out the frequency of a wave from an oscilloscope trace.

timebase setting = 0.2 s/division

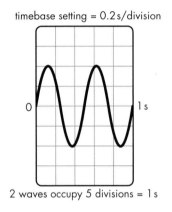

2 waves occupy 5 divisions = 1 s

Properties of waves

The ripple tank is a useful device for showing a range of wave properties. A shallow tank of water is made to ripple with a dipper to produce circular wavefronts, or with a bar to produce straight wavefronts. The direction of travel is perpendicular to the wavefront and can be studied from the projected images of the ripples. Single pulses or continuous waves can be produced. Details are given below. It could be introduced to demonstrate the important relationship between the wavelength, frequency and speed of a wave. Arrange for a straight dipper to produce ripples at a fixed frequency *f*. This will be seen to produce equally spaced ripples across the tank (Figure 2.29). The distance between the ripples is the wavelength λ. These move away from the source at a wavespeed *c*. These quantities could be measured to show that:

$$c = f\lambda$$

The analogy of a runner could be used to explain this important result to pupils. If the runner takes a fixed length stride in the race, then the runner's speed is the length of the pace multiplied by the frequency of the strides.

Figure 2.29
*One arrangement
for showing
waves with a
ripple tank.*

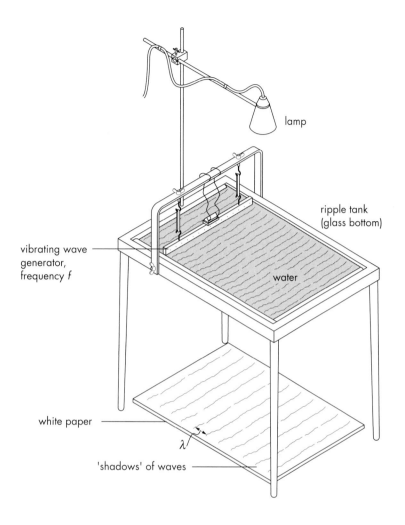

lamp

ripple tank
(glass bottom)

vibrating wave
generator,
frequency *f*

water

white paper

λ

'shadows' of waves

The ripple tank can be used to model the properties of
reflection, refraction, interference and diffraction of waves.
Details can be found in the manuals provided with the ripple
tank kits, or in a range of Nuffield physics publications. A brief
summary is included in the following approach and you will
find some more detailed points in the *Equipment notes*, page 95.

Reflection
A straight barrier can be placed at different angles to a straight
pulse or continuous wave to show the law of reflection. A
concave barrier will show a straight wavefront being brought to
a real focus. A convex barrier will show a straight incident
wavefront diverging after reflection, from a virtual focus behind
the convex surface.

Refraction

This is often quite difficult to see, but it is worth persisting as it provides a key element in the model. Part of the tank is made shallower by immersing a glass or Perspex sheet, and the ripples are observed to move more slowly in this area (they are closer together). When the interface to the shallower water is at an angle to the wavefront, the direction of the front bends towards the normal to the interface (Figure 2.30). Refraction is caused by a change in speed of the wave in different media. When light enters glass from air it is slowed so it bends towards the normal.

Figure 2.30
Refraction of ripples on entering shallower water.

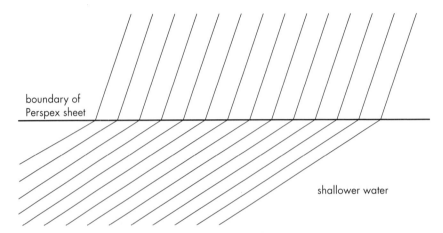

An analogy for refraction is the behaviour of a bike when it runs off the road onto a softer verge. The front wheel hits the verge first and goes slower then the rest of the bicycle so it swings round towards the verge, increasing the danger of an accident. Another analogy is the turning of a troupe of majorettes. Those on the inside take smaller steps (shorter wavelength) but with the same frequency, so they move more slowly. The ranks of majorettes are equivalent to wavefronts.

Interference

This is an important property of waves which distinguishes them from particles. Since waves are simply the temporary displacement of a medium (up or down in ripples), when waves meet the displacements are combined. This results in a process called superposition – one wave is superimposed on another. This can result in a combination from the two waves of crest and crest (increased amplitude, *constructive* interference) or crest and trough (reduced amplitude, *destructive* interference). It can be demonstrated in the ripple tank with a pair of dippers producing identical waves. Where these intersect a characteristic interference pattern of increased and decreased amplitude is produced. It is possible to show light interference, for example through diffraction gratings, where the intersecting light waves are diffracted from the same source (see next paragraph). With white light, the presence of many wavelengths is an added complication which is best not addressed at this level.

Diffraction
Another property characteristic of waves is their ability to 'bend round corners'. This effect is most pronounced at apertures of a width similar to the wavelength. This can easily be shown with a ripple tank, varying aperture and wavelength in the range of a few centimetres. Sound obviously bends round corners; the wavelength of the note middle C is of the same order as the width of a door or window. Light has a very short wavelength, so the necessary aperture is too small to see. Diffraction of light can be demonstrated with a diffraction grating and a laser. The laser beam is deflected into an array of spots across a wide angle. These correspond to the positions of constructive interference, and the gaps to destructive interference.

See the Equipment notes (page 95) for safety considerations.

Shining light through a diffraction grating produces dramatic effects. With monochromatic light (light of a single wavelength, like laser light), bright and dark fringes are seen. With a polychromatic light source (such as a white light), the light is split into its spectrum. Pupils could use a diffraction grating to compare the spectrum of white light with spectra from a range of gas discharge tubes. This simulates the way diffraction spectroscopy is used to analyse the light of stars and so to identify their composition.

Resonance
This is important in the transfer of wave energy. Any system will have favoured frequencies of oscillation. A child's swing needs to be pushed at the same frequency that it is swinging to build up the size (amplitude) of the swing. Push at a different frequency and the swing moves less – its swing is damped. A familiar example is a car which can pick up vibrations from the road or from the oscillations of its engine. At certain engine speeds or on a bumpy road, a part of the car may start to vibrate strongly. The part is said to resonate with the input vibration. Cars are usually built to ensure that these resonances

are damped, for example by the use of absorbent materials. Perhaps the most dramatic example of resonance was the shaking to bits of the Tacoma Narrows suspension bridge in a storm. Video recordings of this event are available.

Seismic waves

Disturbances in the Earth's surface are transmitted through the body of the Earth as seismic waves. There are two types, Primary (P-waves) which are longitudinal and Secondary (S-waves) which are transverse. P-waves are caused by compression and shaking backwards and forwards. They can travel through solids and liquids and travel at 6 km/s. S-waves are caused by sideways or shear forces, and so cannot travel through liquids. They travel at 3 km/s. The waves are detected by seismometers situated around the world, enabling geologists to locate the centre of an earthquake or other disturbance.

Seismic waves are used to find out about the internal structure of the Earth; you could compare this to the ultrasound observation of a developing foetus. Seismic waves are partly reflected and partly refracted at the boundaries between different rocks. The P- and S-waves behave differently at the boundary of the Earth's molten core since S-waves cannot travel through a liquid. This leads to the formation of 'shadow zones' at the surface where no waves are detected. The behaviour of P-waves can be modelled with the set-up shown in Figure 2.31. This represents the layers of the Earth, producing refraction at interfaces and a shadow from the central core. Different patterns of light will be produced if the sizes of the outer and inner 'cores' are altered.

Figure 2.31
Modelling seismic waves.

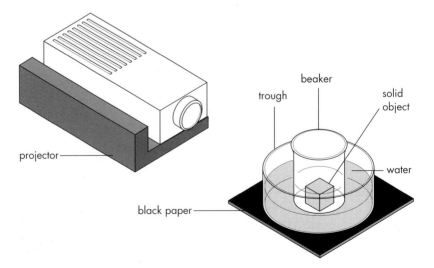

projector

trough

beaker

solid object

water

black paper

◆ *Further activities*

- ◆ A long narrow tank can be used to study the behaviour of water waves. Half fill with water and use a paddle at one end to generate waves of different frequencies and amplitudes. Mixing in a little sawdust can show how the particles of the medium move up and down as the pulse passes through, and that there is also a slight 'rolling' circular motion. This could provide a starting point for an investigation into why and how 'rollers' become 'breakers' in the sea, or a research project into waves as an energy resource (see the first *Enhancement idea* overleaf).
- ◆ The speed of light can be measured in a length of optical fibre using an oscilloscope to display the start of a pulse down the fibre and again when it emerges (Figure 2.32). This will show that it is considerably slower than the speed of light in air or a vacuum.

Figure 2.32
Measuring the speed of light along an optical fibre.

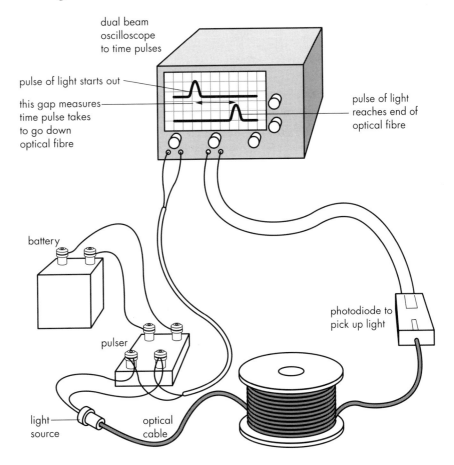

dual beam oscilloscope to time pulses

pulse of light starts out

this gap measures time pulse takes to go down optical fibre

pulse of light reaches end of optical fibre

battery

photodiode to pick up light

pulser

light source

optical cable

◆ The two prongs of a tuning fork generate identical sources of sound. This produces an interference pattern. Strike the fork strongly on a rubber bung. It will provide sufficient loudness for about ten seconds. Rotate it slowly with the tip close to the ear. Four regions of silence should be heard within a 360° rotation. These are points where waves from the two prongs are cancelling each other (destructive interference).

◆ Pupils could devise simple seismometers and consider the different kinds of waves to be detected. Two examples are shown in Figure 2.33 – how could these be made more sensitive? Can they detect heavy traffic nearby?

Figure 2.33
Model seismometers.

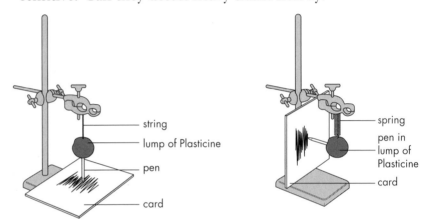

◆ ## *Enhancement ideas*

◆ Water waves offer great potential as an alternative source of energy to fossil fuel. There have been a range of approaches, aimed at extracting the energy of the waves efficiently using equipment that will not be damaged when the waves are very large. Devices tend to be large and therefore expensive and in general have not yet proved economic compared to conventional methods.

◆ Newton's important work on the behaviour of light was done in the belief that light consisted of particles. The work of Thomas Young at the beginning of the 19th century supported the earlier wave ideas of Christian Huyghens. In the 20th century quantum mechanics has shown that light and other electromagnetic radiation can behave both as waves and as particles. Thus light has a wavelength and a frequency and is quantised into packets of energy called photons. The values are, approximately, $\lambda = 10^{-6}$ m, $f = 10^{15}$ Hz and photon energy $E = 10^{-19}$ J.

- Microwave oven doors provide a good example of diffraction effects, or rather their absence. Microwaves have a wavelength of a few centimetres and they are reflected by metal surfaces. Building a metal screen into the door with apertures of less than one centimetre ensures that no microwaves escape. But the doors are transparent because the wavelength of light is very much smaller than the gaps in the screen.
- The regularly spaced tracks on a CD act as a diffraction grating, splitting up white light into a spectrum. Try looking at one in the light of a sodium lamp.

◆ *Equipment notes*

- A simple class **oscilloscope** is best for displaying sound waves, though any type should be suitable. Check the settings of the timebase and the *y*-gain before use, and that the microphone is sensitive enough. To get a steady, reliable trace from an audio signal generator (ASG), connect the oscilloscope to the high impedance terminals of the ASG and a loudspeaker to the low impedance terminals. (The low impedance terminals are usually labelled 8 Ω or 16 Ω.) Most ASGs have settings for triangular and/or square wave outputs. Listening to these while viewing the displays is a good way to show how the shape of the wave affects the quality of the sound – as in different instruments.
- **Sound level meters** are available from many equipment suppliers. The scale is usually marked in dB (A) which is a modification of the decibel scale to take account of the ear's varying sensitivity to sounds of different pitch. As the use of the meter will be comparing sounds, the scale is effectively the same as decibels.
- **Lasers** for school demonstrations should have an output of less than 1 mW – labelled 'Class 2'. By comparison, a filament light bulb is about 3% efficient, so the light output of a 100 W bulb is 3 W. Laser energy is concentrated in a narrow beam, however, so the intensity is proportionately higher. These visible lasers are ideal for showing pupils a 'ray' of light and are essential for easy demonstration of diffraction effects (see Section 2.5).

    Class 3 lasers can damage the eyes, so with pupils under the age of 16 they must only be used for demonstration, and set up so that the direct beam cannot enter anyone's eye. The best arrangement is to mount the laser above eye level, perhaps on a small table on top of the laboratory bench, and to ensure that the diffraction pattern spreads the spots horizontally. Never use a mirror and beware of shiny surfaces such as glass cupboard doors, because of unexpected reflections.

◆ **Light meters** or **light sensors** are available from educational suppliers, to investigate the interaction of light with materials and visibility of surfaces. Use highly reflective tape and road safety clothing to compare with different coloured materials.

◆ A solution of milk of about 1 in 100 dilution will provide good scattering of light with the extent of the scattering being a function of colour. Alternatively, make weak solutions of fluorescein or eosin. Use a strong torch beam or slide projector in blackout conditions for the best effects.

◆ Some rulers and Lego parts are made of fluorescent plastic. Their ends can be seen to glow as light is totally internally reflected along their length, even when bent.

◆ Plastic mirrors lack the high reflectivity of glass or metal ones but they can easily be bent to explore the images, or cut into strips for use in kaleidoscopes or periscopes.

◆ **Ripple tanks** are best purchased from equipment suppliers and need to have the correct illumination for good visibility. The tanks are supplied with a range of accessories to permit the study of wave speed, reflection, refraction, interference and diffraction. There are also guidance notes on how to set up the tank for good results. The following is a brief summary:

- view in near blackout conditions;
- level the tank carefully;
- adjust the depth of the water (about 5 mm) and height of the illumination (about 50 cm) for good contrast;
- for refraction you need an area of very shallow water. Place a Perspex plate in the tank and fill the tank to the same depth as the plate. Then with your fingers drag water over the plate to create a very shallow area;

- use hand-held stroboscopes for viewing, to freeze the pattern when observing wavelengths. (Observe safety precautions if using strobe lighting – avoid frequencies from 4 to 15 Hz.)
- view the ripples as projections onto white paper under the tank, or if as demonstration, onto the ceiling.

All pupils will need help with ripple tank activities; for some classes it would be better to run the activities as demonstrations with pupil assistance. A good solution is to use a tank specially designed to sit on an overhead projector.

References

British Tinnitus Association, 1997: *Don't turn it off, turn it down*. Educational Communications, 50–54 Beak Street, London W1R 3DH (tel: 0171 453 4646; fax: 0171 453 4650; e-mail: edcom@edcom.co.uk).
(Teachers' notes and worksheets for a PSE/science resource on the hazards of loud music.)

SPE, 1996: *Making Sense of Science: Light; Sound* (videotapes). Software Production Enterprises/Channel 4. SPE, The Mansion House, 57 South Lambeth Road, London, SW8 1RH (tel: 0171 793 1882).
(Two of a series of ten videos showing good primary practice on a series of related ideas in a topic theme.)

Other resources

- *Nuffield Primary Science* (Collins Educational, 1993) is a primary course package which makes extensive use of the SPACE project's research findings on children's ideas.
- Various wall charts to aid science teaching are produced by the Pictorial Charts Educational Trust (PCET), 27 Kirchen Road, London W13 0UD.
- The CD-ROM *Light and Sound* is an accessible resource for reinforcing the fundamental concepts, available from Wayland Publishers Ltd, 61 Western Road, Hove, East Sussex BN3 1JD (tel: 01273 722561; fax: 01273 329314).

Background reading

For a summary of research into pupils' misconceptions in science, and alternative frameworks, see:
Driver, R. *et al.*, 1994: *Making Sense of Secondary Science: Research into children's ideas*. Routledge.

For an excellent account of families of musical instruments worldwide, and instructions on how pupils could build and play them, see:
Jackson, A., 1988: *Instruments Around the World*. Addison Wesley Longman.

3 *Forces*

Bob Kibble

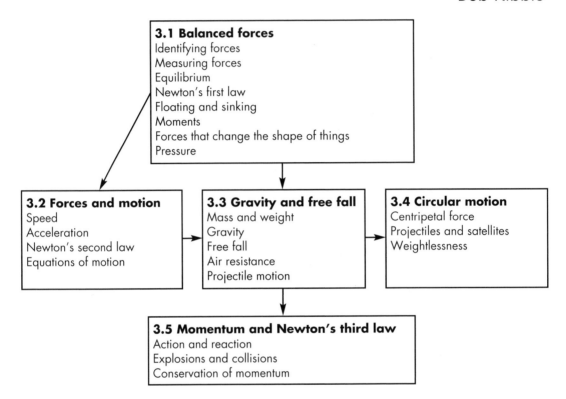

3.1 Balanced forces
Identifying forces
Measuring forces
Equilibrium
Newton's first law
Floating and sinking
Moments
Forces that change the shape of things
Pressure

3.2 Forces and motion
Speed
Acceleration
Newton's second law
Equations of motion

3.3 Gravity and free fall
Mass and weight
Gravity
Free fall
Air resistance
Projectile motion

3.4 Circular motion
Centripetal force
Projectiles and satellites
Weightlessness

3.5 Momentum and Newton's third law
Action and reaction
Explosions and collisions
Conservation of momentum

◆ *Choosing a route*

This chapter starts with a consideration of forces in equilibrium. Why start with balanced forces? Apart from the no-force situation, all equilibrium situations (i.e. no acceleration) involve more than one force, but they are often less complex because they have no changes associated with them. They are by far the most common and helpful in science, engineering and applied maths.

However, an alternative approach would be to start your teaching of forces by focusing on situations where there is a single unbalanced force; that is, start with Section 3.2, which leads naturally to Section 3.3. Section 3.1 could then be inserted once air resistance had been covered. Single forces might relate better to a pupil's experience in primary science (pushing and pulling) but in a world full of friction you are unlikely ever to isolate the pure single force in reality.

3.1 Balanced forces

◆ *Previous knowledge and experience*

The inclusion of forces in the primary science curriculum has been a topic of hot debate in recent years. It is likely that pupils will arrive in secondary school having had very different experiences depending on their school, their home and their teachers. Some, for example, will have used a forcemeter (spring balance or newtonmeter), some will have investigated forces and their effect on moving objects, and some will have tested structures to destruction under loading.

All are likely to associate forces with ideas of pulling, pushing and twisting and most will have measured a force directly, or the effect of a force.

Here is a summary of what may have been experienced:

- forces are brought into play when you push, pull, twist, squeeze or kick something;
- balanced forces can keep things still;
- wheels make things easier to move;
- friction is a force which stops you sliding;
- the force of the wind can make things move;
- when you squeeze a spring, it pushes back on you;
- gravity pulls us down to the ground.

◆ *A teaching sequence*

We should recognise that the idea of force can be difficult to grasp; it is best approached in stages, building on pupils' existing ideas, in the same way as the topic of energy is approached (see Chapter 1).

Forces do not appear in the world fully labelled with their names, magnitudes and directions. So an important skill is to be able to identify forces, give them names, and talk about the object they are acting on and their effects on the object.

Identifying forces

An action, unlike an object, is difficult to visualise. However, forces, being vector quantities (with both magnitude and direction), lend themselves to visualisation through the medium of arrows. Being able to label the forces on a diagram is a skill that is a prerequisite to solving more complex problems later both in physics and applied maths.

In early secondary school pupils might be expected to use arrows and labels to show, for example:

- the pull of a rope on a trolley;
- the drag ('air resistance') of the wind on a train;
- the push of a foot on the floor.

In middle secondary years one might expect pupils to be able to identify and label 'invisible' or non-obvious forces, for example:

- the weight of an apple;
- the drag ('friction') of the ground on a box;
- the upthrust of the water on a boat.

By the end of secondary school a pupil might be expected to identify and label:

- the reaction of the ground on a person;
- the tension in a stretched string;
- the forces during an elastic collision.

Representing forces

It is worth taking care with force arrows. They may compete in diagrams with arrows carrying labels and arrows indicating motion. It is worth trying to distinguish your force arrows from others by using a colour coding – yellow chalk or red whiteboard marker, for example – and by overstating their size, as in Figure 3.1.

Figure 3.1
Representing forces.
a *Emphasise the force by drawing a large, labelled arrow.*
b *An example of a free-body diagram.*

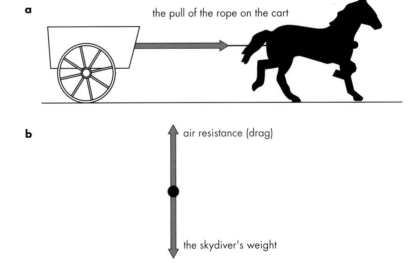

a

the pull of the rope on the cart

b

air resistance (drag)

the skydiver's weight

A problem can arise when drawing a force vector on a diagram which includes, for example, a rope pulling an object. The rope itself is not the force but the force acts along the rope. Pupils are likely to draw a single line to represent both rope and force. It is worth taking a second to illustrate the difference as in Figure 3.1a. A label such as *the pull of the rope on the cart* helps to identify the force.

A complex real-life situation can be reduced and simplified by constructing a free-body diagram. Free-body diagrams focus on the forces and allow the context to fade into the background. They require pupils to think only of the forces *acting on* the body, rather than the forces *exerted by* the body. Figure 3.1b shows an example of a free-body diagram, for a skydiver.

Measuring forces

One of the problems faced by learners is that forces in equilibrium are not always obvious. Tow ropes can be seen as really pulling, but the reaction on a person from the floor on which they are standing is not so easy to identify. In middle secondary years it is helpful to use forcemeters wherever possible to show that forces are present in equilibrium situations. Bathroom scales, kitchen scales and spring balances can all play a part in making forces more obvious in equilibrium situations (Figure 3.2, overleaf).

The idea that the floor can exert a force on an object causes concern to many learners. How can such a passive, dead object like the floor actually do something like force you upwards and do it only when you are standing at that particular place? It appears that the floor somehow knows when you are going to be there and decides to 'turn on' its force at that moment. It is worth anticipating this confusion. By invoking a model of the floor as a molecular structure linked by spring-like forces, the compression in the springs can give a sense of reality to the reaction force. It is similar to the behaviour of a sprung mattress when you sit on it, or a trampoline when you walk on it.

Scale drawings

As an exercise in appreciating the directional (vector) nature of a force, graph paper can be a helpful aid. It helps to establish a scale, e.g. 1 cm represents 10 N, and even a direction framework such as cardinal compass points can be shown. Forces can then be drawn and magnitudes and directions measured. At a later stage, resultants can be drawn and measured. This is particularly helpful when revisiting forces with examination classes or in an advanced course.

Figure 3.2
Ideas for stimulating an understanding of 'hidden' forces.

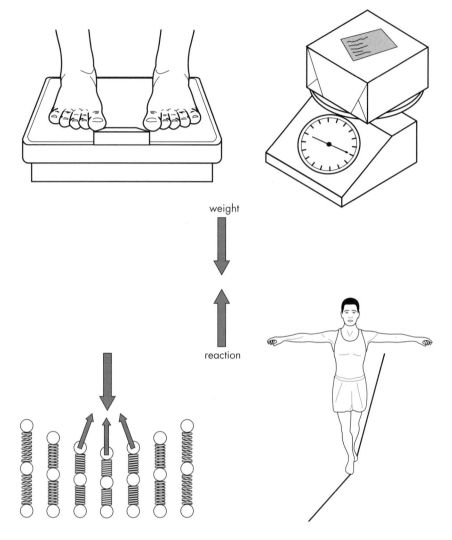

weight

reaction

Equilibrium

Although often reserved for more advanced work, a system of three forces in equilibrium can provide stimulus for a discussion of balanced forces. A good way to set up such a system is to use the white plastic suction hooks available in supermarkets and DIY stores. A pair of these will attach easily to the surface of a wall or, better still, a whiteboard. A light string can be looped onto the hooks and a weight suspended as shown in Figure 3.3a. By attaching forcemeters to the two hooks this situation can be made quantitative. Placing a 360° protractor on an OHP will enable you to project the angles onto the whiteboard for the class to come up and measure (Figure 3.3b). A scale diagram can then be drawn to determine the unknown suspended weight (Figure 3.3c).

Figure 3.3
a A practical arrangement for measuring three forces in equilibrium.
b The three forces.
c A scale drawing (or trigonometry) can be used to find the unknown force.

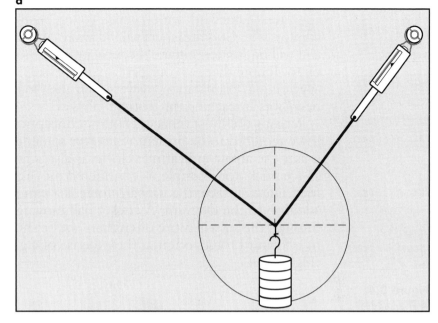

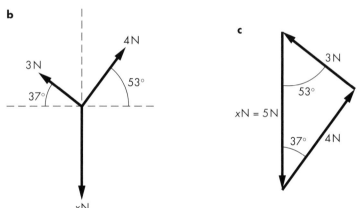

Moving objects in equilibrium

Research has shown that pupils arriving in secondary school bring with them a series of misconceived ideas, many with their origins in pre-school and out-of-school experience. Regarding forces, the most common misconceptions are:

- all moving objects carry with them a force (i.e. force is equivalent to momentum or kinetic energy);
- moving objects stop when their force is used up (i.e. friction is ignored);
- objects at rest have no forces on them;
- force and energy are synonymous and interchangeable.

These four ideas are of course related and form an internally consistent framework within which young people can operate comfortably. Good teachers will be aware of these misconceptions and will not underestimate the tenacity with which they are held on to. Ignore them at your peril – you are likely to find them appearing in examination answers even after you have given your best shots at teaching the 'correct' physics.

Perhaps the most tenacious of misconceptions is that of the force required to be present whenever something is moving. It is as if the arrow indicating velocity is also a force vector. When an ice puck, for example, is considered moving on a completely frictionless surface, the forward force is a natural or obvious addition to the diagram. A cricket ball moving in an arc through the air will more often than not be shown accompanied by a forward force, often at the expense of a gravitational force (Figure 3.4).

Figure 3.4
a The forces acting on a ball moving through the air may be shown incorrectly.
b Ignoring air resistance, the only force really acting on it is gravity.

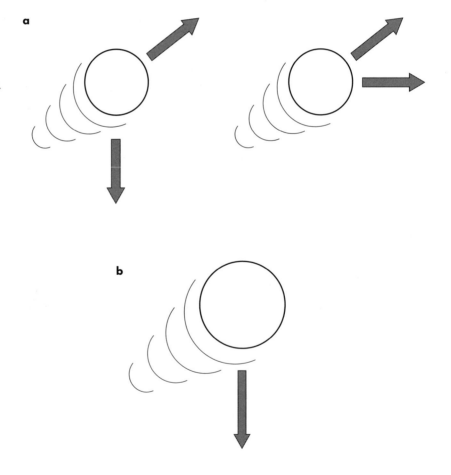

Newton's first law

These days we can draw on examples of motion from space exploration, by NASA probes or perhaps from *Star Trek*, although it is not certain that this will correct the misconceptions discussed above. The deep space probe, moving inexorably into the darkest recesses of the Universe far away from significant gravitational forces, is one of the best examples to illustrate the idea behind Newton's first law of motion – that an object will continue to move at a steady speed in a straight line when no resultant force acts on it. With booster rockets shut down the probe continues to move, unassisted, at constant velocity.

The linear air track is the best terrestrial model of such motion that you are likely to be able to muster in a science laboratory. Making this the focus of a guided discussion is a route forward. By verbalising ideas and bouncing questions around in a group, pupils will be encouraged to rethink their position.

Many teachers make use of the rolling ball model (Figure 3.5). The (frictionless) ball rolls down the curved slope and back up to the position of its initial height. With the longer, sloping track, the ball reaches the same height. Pupils will concede that the ball will always continue to move until it regains its original height. Now consider a horizontal track: the ball will roll on forever, unforced, never reaching the height from which it started.

There is really no practical activity which can establish the 'proof' of the first law. The answer for many pupils lies in chatting and reflecting.

Figure 3.5
A thought experiment to illustrate Newton's first law of motion.

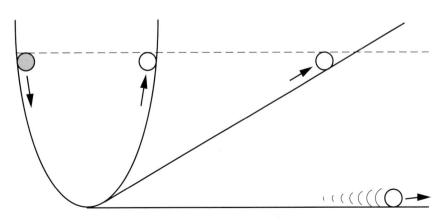

Discussing the first law

'Of course there is a force, Miss, you pushed the trolley to start it moving. That's where it picked up the force.' This type of statement will no doubt emerge from discussion. The initial action, imparting as it does kinetic energy to the moving object, is seen by pupils as 'giving it a force'. Our task is to establish this event as indeed important in supplying energy through a forcing action but that this action is now history. 'The force is no longer with you', to paraphrase a famous *Star Wars* line!

To start a discussion, place a dynamics trolley on a flat runway. Set pupils the challenge of getting the trolley from one end of the runway to the other without touching it. They will do this intuitively, by raising one end of the runway. This can lead into a discussion of the need to overcome friction.

Place a dynamics trolley on a friction-compensated runway – tilt the runway slightly so that a moving trolley continues down the runway at a steady speed. This can provide a good focus for a group discussion about forces and motion. There is no need to take any measurements, simply consider the trolley at rest and then moving at constant velocity. The idea of no acceleration emerges quite naturally from this situation, especially if you deliberately change the angle of slope in your search for the angle for best compensation. The discussion can bring forward ideas about balanced forces but it is best to avoid drawing a free-body diagram. With an advanced class, such a diagram can illustrate forces resolved in directions parallel to and perpendicular to the runway, however this is not a wise path to take with younger learners.

Floating and sinking

Again, here is a topic which occupies pupils from the earliest age at primary school and of course at home in the bath. A standard introductory activity at secondary school is to use forcemeters to weigh objects in and out of water, resulting in conclusions something like: 'Floaters lose all their weight when in water. Sinkers lose only some of their weight.' Home experiments can be encouraged at this point, in the bath, the sink or a bucket. Here are some ideas:

♦ Hold a block of wood and feel the force as you slowly lower it into water. Feel the upward push of the water as you push the block down. How does the force change as the block is pushed further? Does it matter which way the block is held?

♦ Find a ball that floats. Push it under the surface and feel the upthrust from the water. What happens as you push the ball below the surface right to the bottom (of the bath, for example). Does the upthrust go on increasing?

♦ Find an object that sinks in water. Perhaps a (full) tin of baked beans or a glass paperweight. Feel its weight as you lower it into the water. What effect does the upthrust of the water have?

Classifying objects into 'floaters' and 'sinkers' will lead to statements about materials and eventually the density of a material rather than an individual object. As a side issue, this work is also linked to displacement of water and the use of displacement to measure the volume of irregular objects.

To take pupils further towards an appreciation of forces, upthrust and Archimedes' principle, a top-pan balance and container of water is often employed as shown in Figure 3.6. Ask pupils to predict what the top-pan balance will read; then ask them to explain the observation.

Figure 3.6
Exploring forces during floating and sinking.

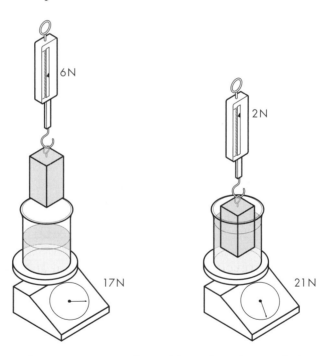

This particular case is a good teaching situation to help appreciate equilibrium. Analysis of the forces acting on the block (2 N + 4 N upwards and 6 N downwards) will allow pupils to appreciate upthrust as a real, measurable force.

Moments

Turning forces are considered only in equilibrium situations in all but the most advanced courses in school. The principle of moments and the confidence to apply the principle to solve problems in one dimension is the main learning outcome.

To simplify the situation, problems most often consider a massless beam, or a beam balanced at its centre of gravity. For practical work, metre rules and special kits can be used to give pupils experience of balancing loads. Alternatively, make a demonstration wooden beam, from a piece of wood perhaps 1.8 m long and 5 cm × 2 cm in cross-section, available from DIY shops for a few pounds. A permanent marker pen will suffice to mark the beam at 0.1 m intervals on either side of the centre of gravity. A wooden block provides an adequate fulcrum. 100 g masses can then be placed on the beam as 1 N forces. The beam is sturdy enough to load an object such as a school bag midway on one side and balance it with an appropriate number of 1 N forces on the other. The sheer size of such a beam lends itself to a controlled demonstration with accompanying discussion and questions. Everyone can see the loads and their positions and so the whole class can tackle problems set by the teacher. (Remember that you are likely to be facing the class and the beam during these demonstrations. Ensure that your diagrams are drawn from the pupils' view.)

Figure 3.7
Using a metre rule and the principle of moments to weigh a bag.

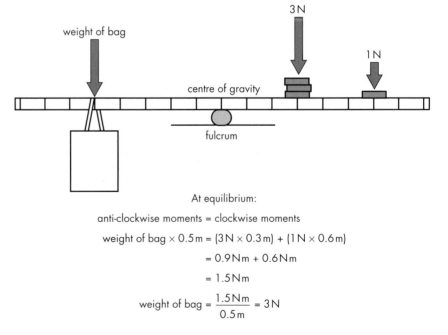

At equilibrium:

anti-clockwise moments = clockwise moments

weight of bag × 0.5 m = (3 N × 0.3 m) + (1 N × 0.6 m)

= 0.9 Nm + 0.6 Nm

= 1.5 Nm

weight of bag = $\dfrac{1.5\,\text{Nm}}{0.5\,\text{m}}$ = 3 N

A typical problem is shown, together with a solution, in Figure 3.7. Note that some confusion may arise because we are ignoring the vertical reaction acting through the fulcrum. In the example shown, we have a total of 7 N acting downwards. Since the beam is in equilibrium, there must be a force of 7 N acting upwards through the fulcrum (neglecting the weight of the beam). You may choose to ignore this force; if you choose to include it, you can restate the requirement for equilibrium, that forces must balance in the vertical direction.

The unit of moment, the newton-metre (Nm) is easily confused with that for work, the joule, since both are obtained by multiplying newtons by metres. In the case of work, force and distance moved are *parallel* to each other; in the case of moments, force and distance are *perpendicular* to each other.

Forces that change the shape of things

Before they reach secondary school, pupils are likely to have stretched springs or elastic bands. However they are unlikely to have plotted accurate force–extension graphs. Such an activity is one of the first times that pupils meet a straight line graph which shows that two quantities are directly proportional. It is worth carrying out such an investigation as it allows a number of key physics skills to be practised: setting up some equipment, accurate measurement, recording results, plotting a graph, drawing a best-fit line, deducing its slope, and evaluation of results. Springs and elastic bands are plentiful and so this is a cheap and easy class practical. (Look for 'expendable steel springs', which are cheap enough to be discarded after they have been stretched past the elastic limit.) Variations on this theme to set as extension work (apologies!) are: two springs in series and parallel combinations (joined end-to-end or side-by-side), elastic bands of different dimensions, clothing elastic, cotton or woollen threads, a pair of braces, a car luggage cord, a party balloon, etc. Careful measurements of an elastic band on loading and unloading will show up both non-linear and non-symmetrical behaviour.

Stretching a long metal wire is often left to advanced classes. The experiment, however, is easy to set up and is worth showing to younger pupils for their sheer disbelief when plastic behaviour sets in. Many will not follow physics at advanced level and this might be the only chance they have to notice permanent plastic deformation in a metal. Choose a wire that will show plastic behaviour at less than 10 N (1 kg) load. Copper is a good choice and a gauge from SWG 28 to SWG 36 (diameter between 0.4 mm and 0.2 mm) will suit smaller loads.

Figure 3.8
Stretching a long copper wire.

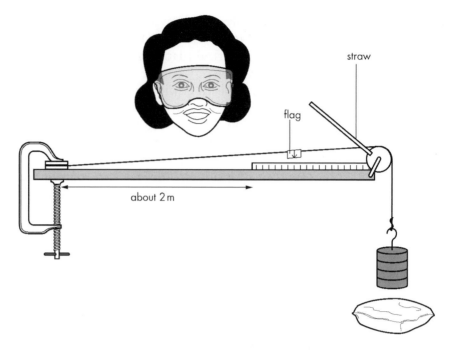

! *Eye protection is essential here, as the wire is likely to break. A cushion on the floor will protect both the floor and the falling masses.*

Figure 3.8 shows two ways to observe the extension of the wire: a flag attached to the wire, and a drinking straw attached to the pulley wheel. At first, with a load of 2 or 3 N, the wire extends elastically, i.e. remove the load and it will return to its original length. This small extension is most easily observed with the straw indicator, since the flag moves only a millimetre or two. Increase the load in steps of 1 N. After the yield point the wire extends more rapidly. This is plastic deformation – the wire has been permanently stretched. It may continue to extend over a period of seconds or minutes, without adding any further load – this is the phenomenon known as creep.

Compressive forces

Often neglected in this work is the behaviour of a material under compression. Compression meters are available from equipment suppliers, but you can get some meaningful results from large pieces of foam loaded with textbooks. Alternatively, a piece of spongy material or a compression spring can be pushed down on a bathroom or kitchen scale to give a reading of the compressing force.

Pressure

Pressure in fluids is included here as a further illustration of forces in equilibrium. (A fluid is any material that will flow, i.e. a liquid or a gas.)

Start by considering the pressure resulting from a force pressing down on a solid object. Discuss situations where you want low pressure (walking on thin ice or boggy ground), and where you want high pressure (cutting with a knife, sticking a pin through fabric). Pupils should recognise that both force and area are significant, and you can define pressure:

$$\text{pressure} = \frac{\text{force}}{\text{area}}$$ (unit: newton per square metre, N/m^2, or pascal, Pa)

Pupils need to meet the concept of pressure in a gas and a liquid. Pressure in a gas is often explained via a particle model and collisions. Pressure in a liquid lends itself more to the idea of the weight of liquid pushing down on a surface. For solids, it might be argued that the term stress is perhaps more appropriate but as pressure is required by curriculum guidelines it is best keep to this term. For all three cases it is desirable that pupils have the capability to use the expression above.

There are plenty of examples of situations where this expression can be applied, including hydraulic jacks, braking systems, connected syringes, snow shoes, and columns of water. A good example is to estimate the pressure exerted by an elephant and by a person wearing stiletto heels.

When introducing the idea of pressure in a liquid, it helps to think of the weight of liquid pushing down on an object. Pupils will have come across the fact that, the deeper you go, the greater the pressure. (This is familiar from diving, submarines, etc.) It may not be obvious that pressure in a fluid doesn't simply push down from above, it pushes from all directions.

You might extend this to develop an explanation of upthrust on a body in a fluid (Figure 3.9, overleaf). The lower surface experiences a greater pressure than the upper surface, and so there is a net upward force.

Figure 3.9
The increase in pressure with depth is used to explain upthrust.

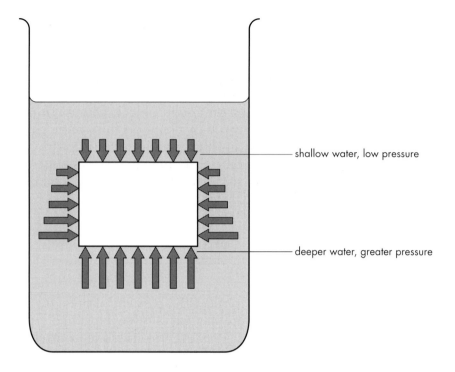

shallow water, low pressure

deeper water, greater pressure

Air pressure

One of the most dramatic illustrations of air pressure is the water barometer. An 11 m length of clear plastic tubing will make a good water barometer. Submerge the open tube in a bucket of water at the bottom of a stairwell, taking care to expel all the air from the tube. Seal the other end with a clip and raise it up the stairwell to a height of about 11 m. Tie some string to the tube to help lift it up. Emphasise the fact that the atmospheric pressure pushing down on the surface of the water in the bucket is pushing the water up the tube. The top 0.7 m or so of the tube is empty – it contains a vacuum.

Water is roughly 800 times denser than air. This suggests that there is 800×10 m $= 8$ km of air pushing down to support 10 m of water, and this is a fair estimate of the height of the atmosphere. The summit of Mount Everest is about 9 km high, enough to pose severe breathing problems to mountaineers (and to cap it all, they can't even brew a decent cup of tea, water boiling at well below 100°C at the top). Of course it is worth pointing out that the atmosphere doesn't suddenly end. Polar orbiting satellites might typically be placed at a height of about 500 km. Even they suffer energy loss through friction with the atmosphere at that height, thin though it might be.

A word about units. Air pressure is quoted in a number of units:

- atmospheric pressure at sea level is roughly 10^5 N/m^2, or 10^5 Pa;
- this is the same as 1 atmosphere or 1 bar or 1000 millibars (mb);
- traditionally, atmospheric pressure was measured using a mercury barometer, giving a typical reading of 760 mm Hg;
- pupils may also come across atmospheric pressure quoted as 15 pounds per square inch (psi). Car tyre pressures are typically 30 psi.

Using a barometer to measure mains gas pressure is an easy classroom task. Use a simple plastic U-tube containing coloured water (add some ink) and mounted on a board. A difference of water height of about 20 cm is to be expected from a laboratory gas tap. This gives the gas pressure as 20 cm of water above atmospheric, i.e. about 10.3 m + 0.2 m. (It must be above atmospheric pressure of course, otherwise it would never emerge when the tap is turned on!)

Some pressure sensors can be connected to computers or dataloggers. They will do the job of monitoring mains gas pressure and much more. Such a sensor can be used to probe pressure variations with depth if the probe is submerged below the surface in a water-filled measuring cylinder.

Brownian motion and the kinetic model

It is useful to develop pupils' understanding of pressure in a fluid in terms of their picture of the microscopic nature of matter. Pressure in a fluid is a result of collisions between fast-moving particles. Two activities are relevant here.

Start with the smoke cell experiment to show Brownian motion. You will need to set this up for your pupils.

1. Connect the smoke cell in its holder to a 12 V lamp supply. Place it on the microscope table.
2. Focus the (low power) microscope on the inside edge of the smoke cell. Slide the cell out, ready to receive the smoke.
3. Light one end of a paper drinking straw. Allow it to burn for a few seconds. Hold it tilted at 45° to allow smoke to drift up inside it.

4. Tip this end into the cell, so that some smoke enters the cell. Cover the cell quickly with a cover slip. Put the straw out safely.

5. Slide the assembly under the microscope and observe.

You need to explain to pupils what they should expect to see. Tiny specks of light move around against a black background. These are illuminated smoke particles. They will drift in and out of focus, moved by convection currents. Pupils may also notice a jerky motion, as the smoke particles are buffeted by air molecules.

Now you can move on to a motorised model of particles in motion, available from equipment suppliers. In this, a 1 cm diameter polystyrene ball represents the smoke particle whilst lead shot represents the 'invisible' air molecules. The base of the apparatus vibrates, causing the ball and shot to move around. Viewed from a distance, perhaps across the laboratory, observers see only the polystyrene ball performing its familiar zig-zag random motion.

◆ *Further activities*

- ◆ Toys often store energy using spring-like mechanisms. Wind-up flying insects and cars are easy to find in toy shops or catalogues. They can be the subject of much fun and learning, as can the jack-in-the-box suction pad jumper. The work done in compressing the spring of such a jumper can be compared with the output potential energy and provide a simple measuring and calculating fun challenge for pupils.

- ◆ Extend work on moments of forces by using a demonstration beam with a displaced fulcrum to find the beam's weight.

3.2 Forces and motion

◆ *Previous knowledge and experience*

Pupils are likely to have timed moving objects, and they may have calculated their speeds. They will have investigated the effects of friction and air resistance. They may also have investigated the movement of cars or trolleys on sloping ramps, although the nature of the force on a sloping runway is unlikely to be apparent to many.

◆ *A teaching sequence*

Before embarking on a consideration of unbalanced forces and their effects ('dynamics'), pupils need a firm grounding in ways of describing motion ('kinematics').

Speed

Start with some practical measurements of speed. This should involve measurements of distance and time. It might involve trolleys or toy cars in the laboratory, or pupils running or cycling in the school grounds. It could also involve monitoring cars on a nearby road, but you will need to be absolutely sure that you can guarantee your pupils' safety. A stretch of road visible from within the school grounds is best. Follow school policy if you take a group off-site and ensure pupils do not interfere with traffic.

Now you can encourage discussion about speed, changing speed and average speed. Consider data from an athletics sprint: e.g. 100 m in 10 s gives an *average* speed of 10 m/s. A rough graph will help to convey the idea that the sprinter's *maximum* speed must be greater than 10 m/s (Figure 3.10).

Figure 3.10
A sprinter's average speed is less than the maximum speed.

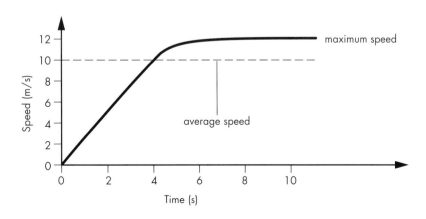

115

You might wish to extend the scale of such motion to that of a train journey from London to Edinburgh (600 km in 4 h). The train has at least one stop during which the clock ticks by but the speed remains zero. The maximum speed then must exceed the average speed of 150 km/h.

You may wish to introduce the term *velocity*, i.e. speed in a given direction. For most purposes at this stage, this distinction is unnecessary.

Acceleration

Now the concept of acceleration needs to be established and distinguished from speed. One approach to this is via graphs of speed against time. Most pupils will have heard that cars can 'do 0 to 60 in ten seconds' – ask them to find some car advertisements with such statements. Here is a place to start. To avoid mph confusion you could boldly introduce the statement that 'a car can 'do' 0 to 30 m/s in ten seconds'. This car could be described as a '3 m/s every second car' – Figure 3.11. (You may wish to try for a halfway-house and mix units such as '5 mph per second' before moving on to m/s per second.)

The target could be to establish, through sounds and physical jerks, that we were talking here about a forceful change of speed – the 'vrooooom' effect. Pupils should also appreciate that the sloping line of the graph indicates that speed is changing.

Figure 3.11
One approach to appreciating acceleration.

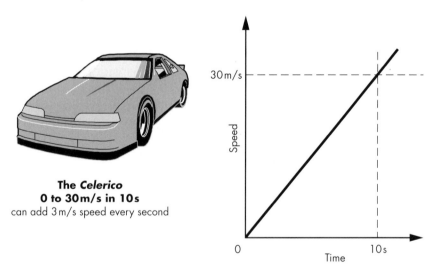

The *Celerico*
0 to 30m/s in 10s
can add 3m/s speed every second

Pupils could be challenged to sketch their own graphs to describe other cars which went from 0 to 40 m/s in 8 seconds, for example. A number of these can be set as a task to reinforce the point. Encourage sketches of the cars themselves to help spice up a potentially dry piece of teaching. A drag-racing video or car magazines can also help to set the scene. Watch out for the mph v. m/s confusion. A conversion graph or table would be handy so that any mph information can be changed directly to SI units (20 mph is about 9 m/s).

Unit of acceleration

Eventually you will want pupils to quote accelerations as m/s^2. Beware of this formulation, however, as it encourages a misunderstanding. Many learners interpret this unit as giving information about a distance travelled – the initial 'm' causes this confusion. Hence an object with an acceleration of $20 \ m/s^2$ is interpreted as something which 'goes 20 m in a (square) second' – whatever a square second might be! You might wish to confront this misunderstanding head on. Be prepared to remind pupils of the origin of the unit: the object gains an extra speed of 20 m/s each second. Later, when discussing falling objects, students will tell you that the acceleration due to gravity is $9.8 \ m/s^2$. When asked how far an object will fall from rest in one second, many will automatically offer 9.8 m as an answer. Try it!

Ticker-timers

In the past, ticker-timers would have been used to establish the relationship between force, mass and acceleration. Gone are the days of endless metres of ticker tape and dotty madness. However, you might wish to use ticker-timers in some simple activities to look at the patterns of dots produced by trolleys or cars moving at steady speed or accelerating.

Using ICT

As part of this development of pupils' ideas of speed and acceleration, they should have experience with electronic timing devices. Typically, these use light gates or infra-red beams to detect a passing object. An 'interrupt card' attached to the moving object breaks the beam; the software is programmed with the dimensions of the card so that it can calculate (average) speed and acceleration (Figure 3.12, overleaf).

There are many different software and timer combinations (see the *Equipment notes,* page 138). It would be inappropriate to focus on any one here; you will have to discover what equipment is available to you. Then you can build a series of activities which run parallel to the activities and discussions outlined above.

Figure 3.12
*Various
arrangements of
light gates and
interrupt cards
allow pupils to
measure speed
and acceleration.
a Gates 1 and 2
detect the leading
edge of the card.
The computer
calculates the
average speed
between the
gates.
b The gate
detects the
leading and
trailing edges of
the card. The
computer
calculates the
average speed at
the gate.
c The gate detects
the four edges of
the card. The
computer
calculates two
speeds, then
calculates the
acceleration from
the change in
speed.*

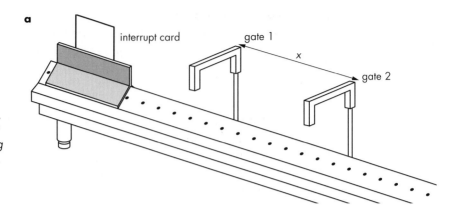

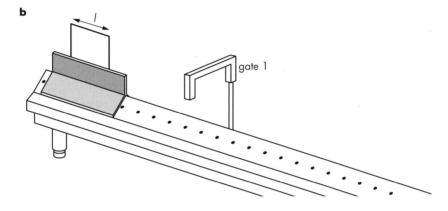

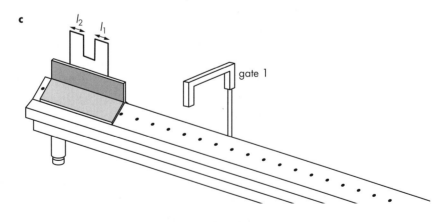

Newton's second law

This is an area where the technology of light gates and timing software comes into its own. The computer does the analysis for you, so that you are free to focus on the interpretation.

Use a trolley on a friction-compensated runway, or a glider on a linear air track. You need to accelerate the trolley or glider with a clearly visible constant force. Perhaps the simplest way is to pull it with a uniformly extended elastic thread or a forcemeter. Results will tend to be qualitative using such a method but you can, for example, demonstrate an approximately proportional relationship by doubling the mass of the trolley.

Figure 3.13
One way to show the relationship between force and acceleration.

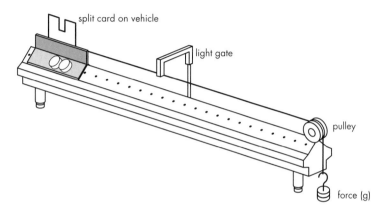

Now set up a trolley or glider pulled by a string over a pulley attached to falling weights, as shown in Figure 3.13. Take care to establish in the pupils' minds that the total mass being accelerated is the mass of the vehicle plus the falling weights. To ensure variables are controlled it helps to ensure that the total mass accelerated is constant. This can be achieved by having a set of, say, six 100 g masses attached to the trolley and transferring them one by one from the trolley to the falling hook. This will give six different values for the accelerating force (each 100 g mass provides a force of 1 N) and hence six accelerations, yet keep the accelerated mass constant. The slope of the force–acceleration graph gives the total mass, m, of the system:

$$\text{force} = \text{mass} \times \text{acceleration} \qquad F = m \times a$$

(Getting pupils to do the experiments and to teach their peers is a good way to ensure that you don't dominate the scene. A keen trio captured before the lesson, perhaps at lunch time, can play with the kit, test it for you and then demonstrate it to the class in a style and a language which might convey more convincing messages than you are able to.)

Inertia

The relationship *force/acceleration = constant* allows you to introduce the concept of *inertia*, defined either as the reluctance of a body to be accelerated or as the result of calculating F/a. Inertia is a difficult idea for anyone learning physics. One approach is to select an object and list its properties, e.g. it has a shape, a colour, a texture, a smell, a weight. Its inertia is simply another property, the one which tells you how difficult it will be to accelerate the object. The term inertia is convenient to use when explaining, for example, why a slack tow rope might break when suddenly tightened by the towing truck. It is a sophisticated concept which you may wish to delay until pupils are mature enough to appreciate it; however, by introducing it now you can revisit it in a short while when dealing with momentum.

Equations of motion

For pupils at this level it is best to avoid introducing equations as if from thin air. Word definitions, accompanied by word equations and graphs (Figure 3.14), are a more user-friendly route.

Figure 3.14
*Speed–time graphs for **a** steady speed, and **b** uniform acceleration. The shaded areas represent the distances travelled.*

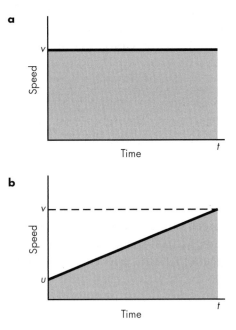

We will now consider, in turn, the equations that are most likely to find a place in this topic.

$$\text{average speed} = \frac{\text{distance travelled}}{\text{time taken}} \qquad v = \frac{d}{t}$$

Substituting symbols for quantities brings its own problems. Apart from their abstract nature there is the issue of which symbol to choose. Should speed be *s* or *v*? Most teachers choose *v* and leave *s* for distance travelled. Others will save *s* for displacement, introduced at Advanced level, and use *d* for distance travelled, as here. You will probably find yourself looking towards your exam board for advice as you will not wish to confuse pupils.

A variation on this equation which is very helpful is:

distance travelled = average speed × time taken

$$d = \frac{(u + v)}{2} \times t$$

where we have used *u* and *v* for initial and final speeds. This assumes uniform acceleration (i.e. constant or steady acceleration), shown on the graph in Figure 3.14b as a straight line. Almost all accelerations considered at GCSE, Standard Grade and equivalent courses are considered to be uniform, the most common exception being the increasing deceleration of a body experiencing drag.

$$\text{acceleration} = \frac{\text{change in speed}}{\text{time taken}} \qquad a = \frac{(v - u)}{t}$$

This can be rewritten in a perhaps more familiar (and more easily memorised) form:

$$v = u + at$$

Note at this stage that this form also has a helpful word equation to support it. This is an equation relating speeds. As such it is helpful to refer to the term *at* as the extra speed added due to acceleration. Hence:

final speed = initial speed + extra speed added

Worked example

Here is a simple example to illustrate these equations in use. The equations of motion appear again in Section 3.3, where we will consider motion under gravity.

A train moving at 20 m/s has its brakes applied for a period of 4 s to bring its speed down to 8 m/s.

a) Assuming the deceleration is uniform, calculate how far the train travels while braking.

Step 1: Write down what you know, and what you want to know:

$$u = 20 \text{ m/s} \qquad v = 8 \text{ m/s} \qquad t = 4 \text{ s} \qquad d = ?$$

Step 2: Select the equation that relates these quantities:

$$d = \frac{(u + v)}{2} \times t$$

Step 3: Substitute the known values and solve the equation:

$$d = \frac{(20 + 8)}{2} \times 4 = 56 \text{ m}$$

b) Calculate the acceleration of the train.

Step 1: $u = 20 \text{ m/s} \qquad v = 8 \text{ m/s} \qquad t = 4 \text{ s} \qquad a = ?$

Step 2: $a = \frac{(v - u)}{t}$

Step 3: $a = \frac{(8 - 20)}{4} = -3 \text{ m/s}^2$

(The minus sign indicates negative acceleration, i.e. deceleration, or slowing down.)

Find time to speak with your colleagues who teach maths. Share your work on equations and the terminology you use. For example, do you 'rearrange' or 'transform' equations? Do you 'change the side, change the sign'? Maths and physics teams seem rarely to have the chance to share their work but it is well worth it if you can.

◆ *Further activities*

♦ Use a motion sensor to investigate motion. This uses a 'radar-ranging' principle to detect the position of a moving object. Software then produces a real-time graph on a computer screen. Pupils learn a lot if they themselves are the moving object.

♦ Use data from a train timetable to investigate the motion of a train. Compare an express train with a stopping train following the same route.

3.3 Gravity and free fall

♦ *Previous knowledge and experience*

Phrases like 'gravity pulls things down' are likely to be all that pupils bring with them from primary school on this topic. Some will have experience of spring balances to weigh things and others may have a notion about gravity within the Solar System. Few of these ideas will have been formalised to give a universal framework.

♦ *A teaching sequence*

Mass and weight

Your pupils may have used the expression *weight* = *mg* implicitly, when using 100 g masses as 1 N weights. So a simple starting point is to talk about 'converting' kilograms to newtons by multiplying by 10. The conversion factor, *g*, should be quoted as 10 N/kg; this quantity is known as the gravitational field strength. (You may see it given as 10 m/s^2, the acceleration due to gravity, but these units do not show its meaning so clearly.) Some syllabuses may require pupils to use *g* = 9.8 N/kg.

Distinguishing between the terms 'mass' and 'weight' seems to be a feature peculiar to physicists and only physicists. Even in other science disciplines the two terms are used interchangeably. Many teachers find that talking about these two words and inviting pupils to use them freely provides a route to their distinction. Statements like 'Imagine you take a 5 kg bag of potatoes to the Moon . . .' are problematic in that pupils have no first-hand experience of being on the Moon (and some may have no experience of a bag of potatoes!). Surprisingly, they do manage to appreciate the problem, especially if accompanied by some theatrical imaginary 'Moon actions', such as slower movement and exaggerated arm and leg gestures as if in space. An alternative is to consider the potatoes being weighed under water. The reading on the weighing machine will change but there will still be 5 kg of potatoes in the bag.

Pupils should understand that *mass* (in kg) tells you something about the number of particles (atoms) in an object; this obviously doesn't change when you go to the Moon. Its *weight* (in N) tells you the pull of gravity on the object, and this depends where you are. (You may wish to introduce your pupils to the idea that mass is a measure of the inertia of an object – see page 120.)

And what of the process of 'weighing'? When we hang the potatoes on a spring balance calibrated in newtons we are 'weighing' the potatoes. What if a balance is calibrated in kilograms? Are we still weighing the potatoes? Strictly speaking the answer is yes, as the balance depends on the effect of gravity on the potatoes to give a reading. The reading would be different on the Moon, despite the fact that the potatoes' mass remained unchanged. The secret is to be clear and consistent without being pedantic.

Gravity

Present in your classroom will be misconceptions about gravity, for example, that heavy objects fall faster, that gravity goes 'up to the sky' or only to the edge of the atmosphere, that the Earth's gravity comes from a magnet in the Earth, that there is no gravity in a vacuum and perhaps no gravity anywhere in space. Gravity may even be associated with the Earth's rotation by some pupils: we would all fall off if ever the Earth were to stop spinning; what's more, the faster the Earth spins, the more likely we are to fly off! They want it both ways.

Gravity is one of the four fundamental forces of nature. It acts between any two bodies with mass. It is not possible to provide an 'explanation' of the origin of gravity which will satisfy pupils – it remains one of the great mysteries of science; however, you can hope to reinforce their understanding of its nature. Perhaps the best way to do this is to deal with gravity in the context of an understanding of the Solar System. The planets are massive; the Sun is more massive still. The gravitational pull of the Sun on the planets is enough to hold them in their orbits. You can extend this to the Earth–Moon system, artificial satellites etc. The paths of spacecraft which fly past several planets are a dramatic example of the effect of gravity.

Free fall

Eventually you will want to explore free fall and determine a value for the acceleration due to gravity, g. There can be few physics teachers who have not taken their lives into their hands and clambered atop a desk to drop a large and a small rubber bung or equivalent. It is surprising how many pupils will predict that the heavy bung will land first as it 'has more force'. Rubber bungs are cheap and there will be plenty for pupils to do this in pairs. Adding a feather or sheet of paper will extend the discussion to bring in the 'hidden' force of air resistance.

Pupils appreciate the story of Galileo dropping a large and small ball from the Leaning Tower of Pisa, although this is probably apocryphal. Air resistance would have played a significant part, resulting in the smaller ball being about a metre or two behind the heavier ball on impact.

There can be few practicals that offer more ways to determine a physical constant than the determination of *g*. Indirect methods include the timing of a simple pendulum. However, direct methods are to be preferred. Here are brief notes on four:

♦ Drop a heavy object attached to a ticker tape and timer. Analyse the dot spacing to find the acceleration. Friction between tape and timer is a problem here.

! *Position a cardboard box to catch the object. Ensure pupils stand clear as it falls.*

♦ Use an electromagnet, trap door and timer. The electromagnet is switched off to release a steel ball and start the timer. The ball falls through the trap door, stopping the timer.
♦ Drop a ball which houses a digital clock. The clock starts on release and stops on impact with the ground.
♦ Use an interrupt card attached to a weight. The card has a rectangular hole cut in it, so that the beam of a light gate is broken twice by the falling card. This method has the advantage that you can measure the acceleration at different heights, to show that *g* is constant.

The second and third methods above measure the time *t* taken for the object to fall through height *h*. To calculate *g*, proceed as follows:

1. Calculate the average speed = h/t.
2. Since the object started from rest, its final speed must have been twice this = $2h/t$.
3. It took time *t* to reach this speed, so acceleration = change in speed/time taken = $2h/t^2$.

Using the equations of motion

Numerical problems involving acceleration by gravity can often be solved using the equations of motion. Care must be taken with signs. A simple convention is to consider upwards as positive, in which case *g* is negative.

<u>Worked example</u>

A ball is thrown upwards with an initial speed of 15 m/s.

a) How long will it take to reach its highest point?

At the highest point, its speed is zero.

Step 1: $u = +15$ m/s $v = 0$ m/s $a = -10$ m/s^2 $t = ?$

Step 2: $v = u + at$, so $t = \dfrac{(v - u)}{a}$

Step 3: $t = \dfrac{(0 - 15)}{-10} = +1.5$ s

b) How long will it be before it reaches the ground again?

When it reaches the ground again, symmetry tells us its speed will be 15 m/s downwards. Symmetry also tells us the answer must be twice that in a), i.e. 3 s. Alternatively, apply the same equation again.

Air resistance

You may wish to explore the effect of air resistance. The 'sycamore fruit' practical is a useful approach (Figure 3.15). This works well if set as a challenge to see who can make the paper propeller that takes the longest to reach the ground.

Figure 3.15
A design for a 'sycamore fruit' helicopter.

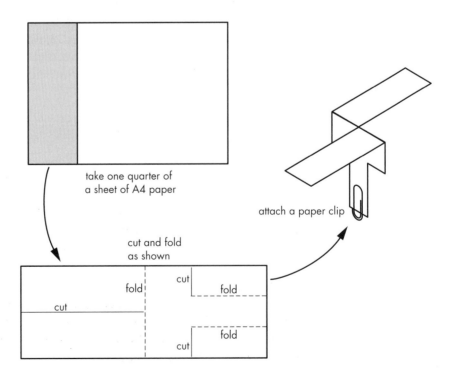

take one quarter of a sheet of A4 paper

attach a paper clip

cut and fold as shown

cut

fold

cut

fold

cut

fold

Terminal velocity

Pupils should appreciate that air resistance (drag) increases with speed. Think about walking in a swimming pool. The faster you try to move, the harder it is. Pupils are familiar with the idea that a parachute prevents the user from falling faster than about 10 m/s. This is their terminal velocity. Free-fall skydivers enjoy falling faster than this (for a short while), and adjust their body shape to control air resistance.

A car is also affected by air resistance, and this is why cars, trains, planes etc. have a top speed. The drag force at top speed equals the forward force provided by the engine.

Projectile motion

A projectile is an object which is projected (thrown or fired) so that it moves through the air. Once it leaves your hand, it has no source of kinetic energy, unlike an aircraft or rocket. Gravity acts on it (downwards, at all times) and perhaps air resistance. Pupils find it difficult to appreciate that there is no forward force acting as the projectile flies through the air (Figure 3.4, page 104).

In the absence of air resistance, a projectile moves equal horizontal distances in equal times, while its vertical motion is accelerated by gravity. This is an important example of a situation where an object's motion can be best understood by dividing it into components in two directions (Figure 3.16, overleaf). You can demonstrate this by rolling a ball diagonally on a sloping board – its path will be parabolic. Because the ball is not moving in an exactly vertical plane, you have 'diluted' the effect of gravity, and the path is easier to see. Pupils can also study multi-flash photographs of projectile motion; most textbooks have examples. Computer simulations are also available.

Giving separate consideration to a projectile's horizontal and vertical motions will allow you to stress the vector nature of velocity.

The 'monkey and hunter' practical helps to emphasise the effect of gravity on a projectile. The 'monkey' is an object such as a tin can, held at ceiling height by an electromagnet. It must be heavy enough to be released when the circuit is broken. The hunter's bullet is a marble fired from a spring-loaded 'gun' (you can use an old dynamics trolley). On firing, the marble breaks an electrical contact in the electromagnet circuit (a flimsy kitchen foil contact is easily broken).

Position a cardboard box to catch the falling 'monkey'. Ensure pupils stand clear of both the flying marble and the falling 'monkey'.

Figure 3.16
Understanding the motion of a projectile. Its path through the air is a parabola.

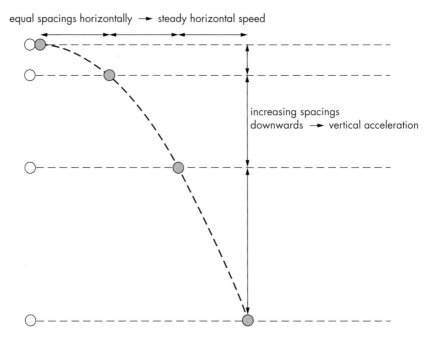

The story is that the wily monkey sees the hunter and decides to drop from its branch as the trigger is pulled. It thinks the bullet will pass over its head; in practice, the two objects are pulled downwards equally by gravity and the marble strikes the falling monkey. This demonstration is worth setting up but it might take your whole lunch time to get it working properly so that the hunter strikes every time! In principle, the gun must be aimed directly at the monkey. Since shooting wild animals is a sensitive issue, you could set the story in the past, or substitute 'archer and apple'.

◆ *Further activities*

You will find more ideas for activities related to gravity in Chapter 5, *Earth in Space*.

◆ *Enhancement ideas*

◆ Take care not to spread the notion that Newton 'discovered gravity'. His great realisation was that the Earth's gravity, the force that pulls a falling apple to the ground, can also account for the movement of the Moon around the Earth. In other words, the Earth's gravity extends far beyond the Earth, and the heavens are governed by the same forces as objects on Earth.

3.4 Circular motion

Up until now, we have mostly considered situations where an object is in equilibrium, or is uniformly accelerated in a straight line. In circular motion, an object is constantly changing its direction of motion, although its speed may not be changing. You may decide to tackle this simply through a consideration of, say, polar and geostationary satellites, thereby bypassing the difficult conceptual physics behind circular motion. A syllabus that emphasises the practical applications of physics might entice you so to do. However, an introduction to circular motion is the tip of an iceberg which might one day allow the learner to probe the deep levels of rotational dynamics, torque, angular momentum, etc., so you may wish to go into this topic in greater depth.

◆ *Previous knowledge and experience*

Pupils may have come across a specific example of circular motion – the orbiting of planets around the Sun – and they may have an appreciation that it is gravity that keeps them on their circular paths.

◆ *A teaching sequence*

It seems intuitively obvious that objects moving in a circle are thrown outwards. The idea of 'centrifugal force' is in everyday parlance. Children experience this on a playground roundabout. They know that water is thrown outwards from clothes in a spin-drier. And they may think that astronauts would fall down to Earth if there wasn't an outward force keeping them up there. These are the misconceptions you have to overcome.

Centripetal force

A good demonstration is to swing a bung attached to a length of string around your head. Have the class grouped to one side of the room and enlarge your circle to a radius of about two or three metres if space will allow. Invite suggestions of what will happen if you release your grip. With a careful countdown you can time your release to occur just as the bung passes your pupils' noses!

The majority of the pupils will predict that the bung will move off in the direction A → C or A → D (Figure 3.17) and will be surprised by the fact that it goes A → B. The natural motion of the bung is to continue in a straight line, in this case tangentially to the circle. (Take care to check that there is not a window or a sensitive piece of apparatus set up in line with the tangential path.)

Figure 3.17
Investigating circular motion (view from above).

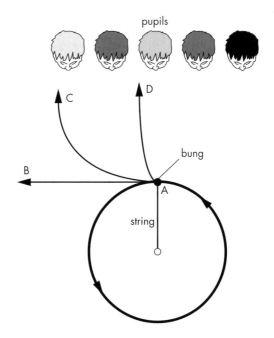

You can move on to discuss this result. On hand to help you is Newton with his first law of motion. Think about an object moving in a straight line. Your approach might be along the lines of: 'How can a moving object be persuaded to depart from a straight line – to take a deviated path towards a circle? The first law suggests that a force would be required to do this. For any object it would be hard to justify a force to the right being responsible for a deviation to the left. The force must be in the direction of deviation ...'. This force is described as a *centripetal* (centre-pointing) force. The term centripetal simply describes the direction of the force. The *cause* of such a force might be tension in a string, gravity, road friction, etc. They can all be considered to be centre-pointing. Many pupils think that there exists a particular force called a centripetal force, rather like a force called friction or a force called weight.

You will need to discuss these ideas and the common centrifugal/centripetal misconceptions with pupils. They are keen to learn about how many people get their physics wrong. Some may do their own mini-research study over the weekend, questioning friends and family about where the forces might be, where the bung might land, etc. In doing this the pupils are brought into the centre of the controversy.

We often use the Moon as an example of an orbiting body under the influence of a centripetal force. You may wish, as an alternative, to refer to Jupiter's moons. The four outermost moons move in the opposite sense to the other twelve. It has been suggested that pupils will be less likely to attribute a forward force to each moon on a diagram where two moons are moving in opposite senses. This is a situation worth exploring as it may help pupils to accept the existence of a single centripetal influence regardless of direction or orbit.

Projectiles and satellites

Newton's thought experiment of the cannon on a hill is another source of material for discussion. If fired with sufficient speed, a cannon ball projected horizontally from a mountain summit will orbit the Earth. Too slow and it hits the ground; too fast and it leaves the Earth (Figure 3.18). Pupils should appreciate that the direction of gravity is always towards the centre of the Earth.

Figure 3.18
Newton's thought experiment.

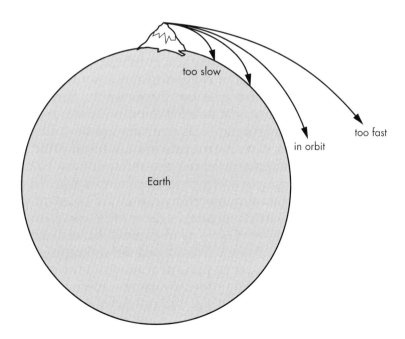

too slow

too fast

in orbit

Earth

Software is available for pupils to explore the effect of changing the initial speed of the cannon ball – for example, the CD-ROM *Multimedia Motion* (see *Other resources*, page 138).

The different functions of satellites in polar orbits (mainly imaging and mapping) and geostationary orbits (mainly communication) can be discussed, the latter with reference to satellite television dishes. The fact that these dishes don't move and hence don't need to track the satellite tells us something about the motion of the satellite relative to Earth. Of course the dish is tracking the satellite by virtue of being fixed to a spinning Earth. The direction of domestic dishes, towards the south in the UK, helps reinforce the fact that the satellites lie above the Earth's equatorial plane. (Which direction would they expect satellite dishes to point in Australia and in central Africa?)

On clear nights it is quite possible to see a number of passing satellites in their polar orbits. To follow these, with binoculars for example, we clearly need to move our eyes across the sky and so track at a faster and clearly detectable rate (typical polar orbit periods being about 90 minutes). Look for satellite predictions in the weather section of some newspapers. Satellites are more easily seen away from town lights; ensure that children are accompanied by an adult if they wish to do some evening observation.

Weightlessness

Mass and weight are difficult enough for pupils to untangle; the idea of weightlessness in an Earth-orbiting satellite can add to the confusion. Pupils will be familiar with images of astronauts floating about in space; the obvious, but incorrect, interpretation is that they are beyond the pull of the Earth's gravity. Astronauts in near-Earth orbit are only a few hundred kilometres above the surface of the Earth, where the Earth's gravity is only a few percent weaker. In the terms of Newton's thought experiment, they are constantly falling towards the Earth's surface; because of its curvature, they never get any closer. One way to express this is to say that 'the pull of the Earth's gravity is 'used up' in keeping them in orbit; none is left over to pull them down to the ground'. If they were travelling more slowly, less force would be needed to keep them orbiting, and the remainder of the pull of gravity would make them spiral down to Earth.

3.5 Momentum and Newton's third law

This final section deals with some ideas which might be thought applicable only to the more able learner. However, there is a growing body of teachers who recognise that the concept of 'momentum' (the 'motion in an object') offers a more natural way of understanding forces and their effects. Perhaps it should be introduced much earlier in the teaching of dynamics.

◆ *Previous knowledge and experience*

Newton's third law of motion is not easy to grasp. The idea of action and reaction has come into everyday parlance, but this is likely to serve only as an impediment to the understanding of the third law.

◆ *A teaching sequence*

Action and reaction

Isaac Newton realised that forces are always created in pairs. You push on the wall and the wall pushes back on you. The Earth's gravity pulls on the Moon and the Moon's gravity pulls on the Earth. If you pull on your boot laces, they pull back on you – that's why you can't pull yourself up by your boot laces.

The Earth–Moon example illustrates the nature of a pair of forces in the sense of the third law:

- the forces are both of the same type (gravitational);
- they act on different objects (Earth and Moon);
- they are equal in size;
- they act in opposite directions.

Confusion can arise with the balanced pairs of forces which may act on an object in equilibrium. This is a very different situation: they may be of different types, and they act on the same body. This distinction is illustrated in Figure 3.19 (overleaf).

Figure 3.19
Pairs of forces.
a *These two objects are in equilibrium; each has two equal and opposite forces acting on it.*
b *Here, the two equal and opposite forces are acting on different bodies.*

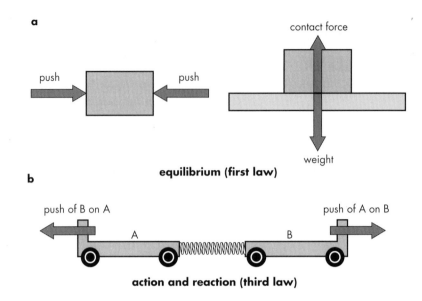

a

push → ← push

contact force

weight

equilibrium (first law)

b

push of B on A

push of A on B

A   B

action and reaction (third law)

A further confusion arises from the use of the word *reaction* to refer to the contact force between two objects. Some teachers avoid this usage, and simply refer to this force as the contact force.

Space exploration provides an engaging context in which to place ideas about action and reaction. For many pupils the first time they will have met such forces will have been when watching a demonstration of a party balloon being blown up and released. In space exploration, from launch to mid-flight adjustments, the same principle applies. Booster rockets eject a gas in one direction. This action is accompanied by a reaction force on the rocket. Because of the prominence given to the launch many pupils will tend to think that the rocket needs something to push against. Even in the first second or so when a launched rocket appears to push against the launchpad, it is still the momentum of ejected gas that provides the 'action'.

Plastic cola bottles, 1.5 or 2 litre, make excellent water rockets. Quarter-filled with water and attached to a foot pump (Figure 3.20a), these rockets will easily clear the roof of a four-storey school block. As a refinement you can attach a simple parachute made from a plastic carrier bag and cotton thread which is housed under a paper nose cone (Figure 3.20b). The cone will remain in place through air resistance on the upward journey, the chute deploying itself on the descent. You can buy a special water rocket kit (*'Rokit'*) if you don't wish to rig up your own, but you will still need to provide your own bottle and foot pump. Foot pumps are preferable to bicycle pumps, which may not be up to the job of repeated launches.

These rockets are very safe but during descent they will reach some speed and any fins could injure someone standing in the way. The only precaution is to keep observers well back.

The reason for discussing motion in space is that we can consider the object free from the effects of friction. Pupils will be familiar with images of astronauts working outside their craft. If they drift away, how can they get back? (Throw a spanner in the opposite direction.) Other 'frictionless' situations include an ice rink or a polished floor.

Figure 3.20
a A lemonade bottle rocket ready for launch.
b Attaching a parachute and nose cone.

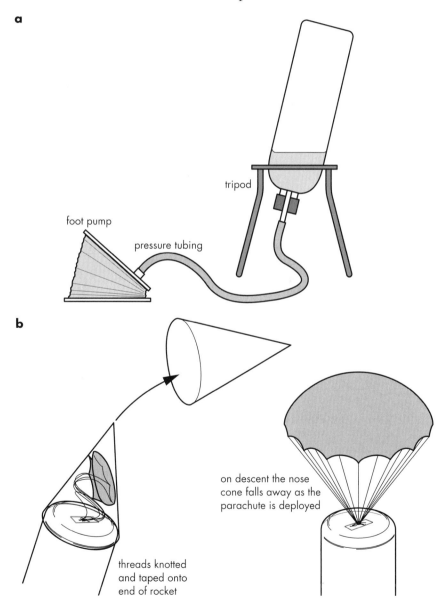

a

tripod

foot pump

pressure tubing

b

on descent the nose cone falls away as the parachute is deployed

threads knotted and taped onto end of rocket

Explosions and collisions

A controlled 'explosion' or a collision can be an excellent focus for a discussion about forces and momentum in a novel situation.

Start with a single trolley with its spring-plunger released. Push it towards a second trolley. The spring ensures a springy (elastic) collision. The first trolley halts and the second moves off at the speed of the first.

Now try an explosion with two trolleys or two air-track vehicles with repelling magnets. Two dynamics trolleys are preferable, as you can easily add another to double the mass (as shown in Figure 3.21). Tap the peg and the spring-plunger pushes the trolleys apart. On a level surface friction will act on both trolleys but the effect is clear enough for qualitative commentary.

Figure 3.21
Ready for a demonstration explosion. Which trolley will move faster?

release

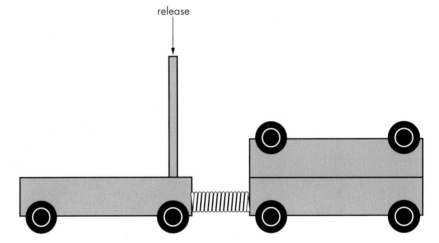

Pupils could try this out for themselves and be asked to come up with some conclusions of their own. Then you could set up an air track or trolleys with light gates to test their ideas. You should be able to show that, when two individual trolleys explode apart, they move off with equal and opposite velocities. A double trolley moves with half the speed of a single one. It is a short step from this statement to an appreciation that *mass × velocity* is the same for both but in opposite directions. The definition of momentum has arisen naturally from these observations. You can then ask your pupils to explain the simple trolley collision demonstration in terms of momentum transfer.

Next, the forces acting and their symmetry is the target for discussion; both trolleys are pushed by equal forces for equal times but in opposite directions. (For trolleys pushed by a released spring the force is of course not constant but should this be raised then you can refer to the average force acting.)

Conservation of momentum

Momentum is always conserved: there is nothing it can be transformed into. (This is unlike kinetic energy, which can be transformed into other forms of energy.) But isn't momentum created out of nothing when the trolleys explode apart? Momentum can appear to pupils to come from nowhere. Pupils must understand the vector nature of momentum to appreciate that here are two momenta, equal and opposite, their total remaining zero at all times. Similar, but perhaps more sophisticated, problems arise when considering a piece of Plasticine thrown at a wall. The momentum appears to disappear on impact. The concept of the Plasticine + Earth system is needed to come to your rescue on this problem. A falling object is another case where momentum appears to come from nowhere; in fact, the Earth moves upwards microscopically to meet it, pulled by the object's gravity. Most pupils enjoy hearing stories of the Earth moving in such discussions.

Momentum and kinetic energy

Learners and teachers grappling with momentum and kinetic energy for the first time together are bound to struggle with the difference between these two ideas. Both involve mass and velocity, but there is no easy mental picture that will help to distinguish them. It may help to emphasise the vector nature of momentum. The direction of motion can be considered as positive or negative, hence positive and negative velocities and momenta. In calculating kinetic energy $(=\frac{1}{2}mv^2)$ we have to square the velocity, so all negative quantities become positive. This may help some pupils to realise that kinetic energy is always positive, regardless of the direction of motion. 'Kinetic energy takes no notice of direction.'

♦ *Further activities*

- ♦ Experimenting with 'frictionless' pupils can be fun. Arrange for two or more pupils to wear roller skates, or stand on roller boards. Then one can push another to experience recoil. Similarly, one can throw a heavy object to the other. They should be able to explain their observations in terms of forces and changes in momentum.

◆ *Equipment notes*

 ◆ The range of ICT equipment for datalogging activities in science continues to expand, making it impossible to cover every system for recording speed and velocity data. For a useful guide to equipment and activities, see *IT Activities for Science 11–14* and *IT Activities for Science 14–16*, Carol Chapman and John Lewis (Heinemann, 1998 and 1999). These include detailed instructions for using equipment supplied by Philip Harris, Griffin & George and Data Harvest.

◆ *Other resources*

◆ Two titles in the *Nature of Science* series, *The Big Squeeze* and *Stars and Forces* (ASE, 1989), deal with historical aspects of pressure and gravity respectively, and are suitable for older secondary pupils.

 ◆ A CD-ROM called *Multimedia Motion* allows you to show and analyse many different types of motion, including various vehicles, spacecraft and athletes; available from Cambridge Science Media, 354 Mill Road, Cambridge, CB1 3NN.

 ◆ Another CD-ROM is *Fun Physics*, which allows pupils to model motion in a variety of situations; available from TAG Developments, 19 High Street, Gravesend, Kent, DA11 0BA.

◆ Look out for the historic photographs by Eadweard Muybridge, showing people, animals and birds in motion. These helped people to understand animal locomotion.

 Web sites

For activities on flight, look at the *Flights of Inspiration* pages on the web site of the Science Museum: **www.nmsi.ac.uk**

Visits

For practical experience of many activities involving forces, visit one of the many hands-on science centres such as Techniquest in Cardiff, Satrosphere in Aberdeen, or the Bristol Exploratory.

Background reading

Many people made important contributions to our understanding of forces and motion. You can find readable accounts of the work of Galileo, Newton and Einstein in:
Bragg, M., 1998: *On Giants' Shoulders*. Hodder and Stoughton.

For a detailed discussion of the mechanics of flight using little more than formulae accessible to a bright 16-year-old, see:
Tennekes, Henk, 1997: *The Simple Science of Flight*. MIT Press.

4 *Electricity and magnetism*

Joan Solomon

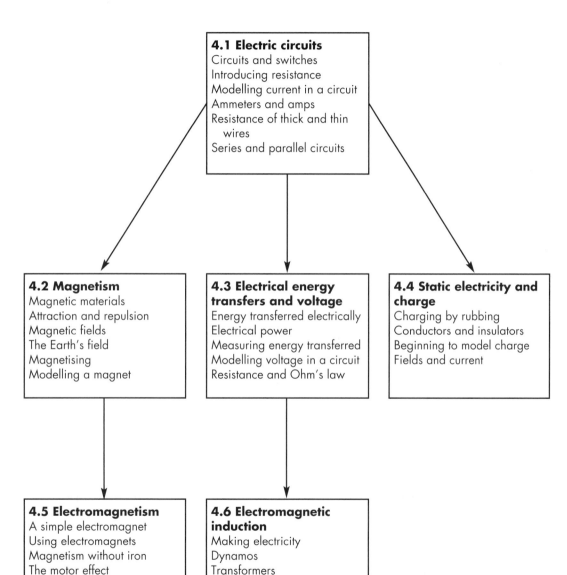

4.1 Electric circuits
Circuits and switches
Introducing resistance
Modelling current in a circuit
Ammeters and amps
Resistance of thick and thin
   wires
Series and parallel circuits

4.2 Magnetism
Magnetic materials
Attraction and repulsion
Magnetic fields
The Earth's field
Magnetising
Modelling a magnet

4.3 Electrical energy transfers and voltage
Energy transferred electrically
Electrical power
Measuring energy transferred
Modelling voltage in a circuit
Resistance and Ohm's law

4.4 Static electricity and charge
Charging by rubbing
Conductors and insulators
Beginning to model charge
Fields and current

4.5 Electromagnetism
A simple electromagnet
Using electromagnets
Magnetism without iron
The motor effect
Motors

4.6 Electromagnetic induction
Making electricity
Dynamos
Transformers

◆ *Choosing a route*

Teachers have their own ideas about what is more or less fundamental for guiding the order in which they teach. The path chosen here accords with the author's experience of the difficulty children find with the different parts of this subject. However, you are entirely free to choose your own path. Let a thousand flowers bloom!

As they come into the unfamiliar surroundings of secondary school it is all too easy for the pupils to seem to 'forget' what they have learnt. Secondary teachers often claim that their new pupils know nothing, but this is not true. Unfamiliar surroundings strongly influence the ease with which we recall what we know. (Remember that we often 'forget' what we had intended to get from downstairs when we were upstairs and need to go back upstairs to recall it!) It is far better to remind the pupils of work they have already done, than to test them, or to teach it to them all over again. Many pupils complain that, in the first years of secondary school, teachers keep teaching them what they already know.

The rationale for choosing where to start begins with what the youngest pupils find easiest. From the earliest years in primary school, they have learnt:

- how to make simple circuits which light up a bulb;
- the difference between conductors and insulators;
- that magnets attract some metals (iron and steel).

Most of them are best at magnetism and simple circuits, so that is where it seems good to start, giving them credit for what they can already do. However, you must make the work different and new lest it becomes boring.

The second criterion is to keep apart things that the pupils often mix up (e.g. magnetism and electrostatics, or current and voltage.) These should be taught separately and as far apart in time as can be managed.

Finally, if at all possible, pupils should, in each of Sections 4.1 to 4.6, experience the five important physics skills: *observing*, *measuring*, *modelling* an abstract concept, *predicting* or *explaining* from a model, and *carrying out* an investigation.

4.1 Electric circuits

◆ *Previous knowledge and experience*

In primary school, pupils will have learnt to construct working circuits which include such devices as bulbs, batteries and switches. They will also have learnt to distinguish conductors and insulators.

◆ *A teaching sequence*

Circuits and switches

Start with a challenge to make a circuit of their own. Can they design a circuit with a switch, for example to turn on a light or ring a bell if a thief steps on the doormat? Let them design the switch on paper first. Then they should be able to make a list of the simple materials and equipment they want. Since they have probably already experienced this type of activity in primary school, they could do part of it for homework.

If they seem very good, try asking them to design a circuit with three switches: one beside Mary's bed, one beside John's bed in another room, one for Mum to turn off both lights when she thinks they should stop reading and go to sleep (Figure 4.1).

Figure 4.1
A multiple switching circuit.

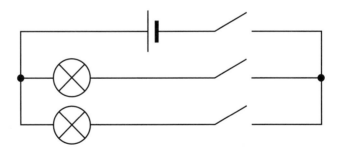

Circuit symbols

This activity will have revealed which of your pupils are familiar with conventional circuit symbols. For those pupils who draw pictorial representations of their circuits, you can take the opportunity to introduce some conventional symbols. As this topic develops, you can extend their repertoire by introducing more symbols. Figure 4.2 (overleaf) shows some basic symbols. For a detailed treatment of circuit symbols, see *Signs, symbols and systematics: the ASE companion to 5–16 science* (ASE, 1995).

ELECTRICITY AND MAGNETISM

Figure 4.2
Circuit symbols.

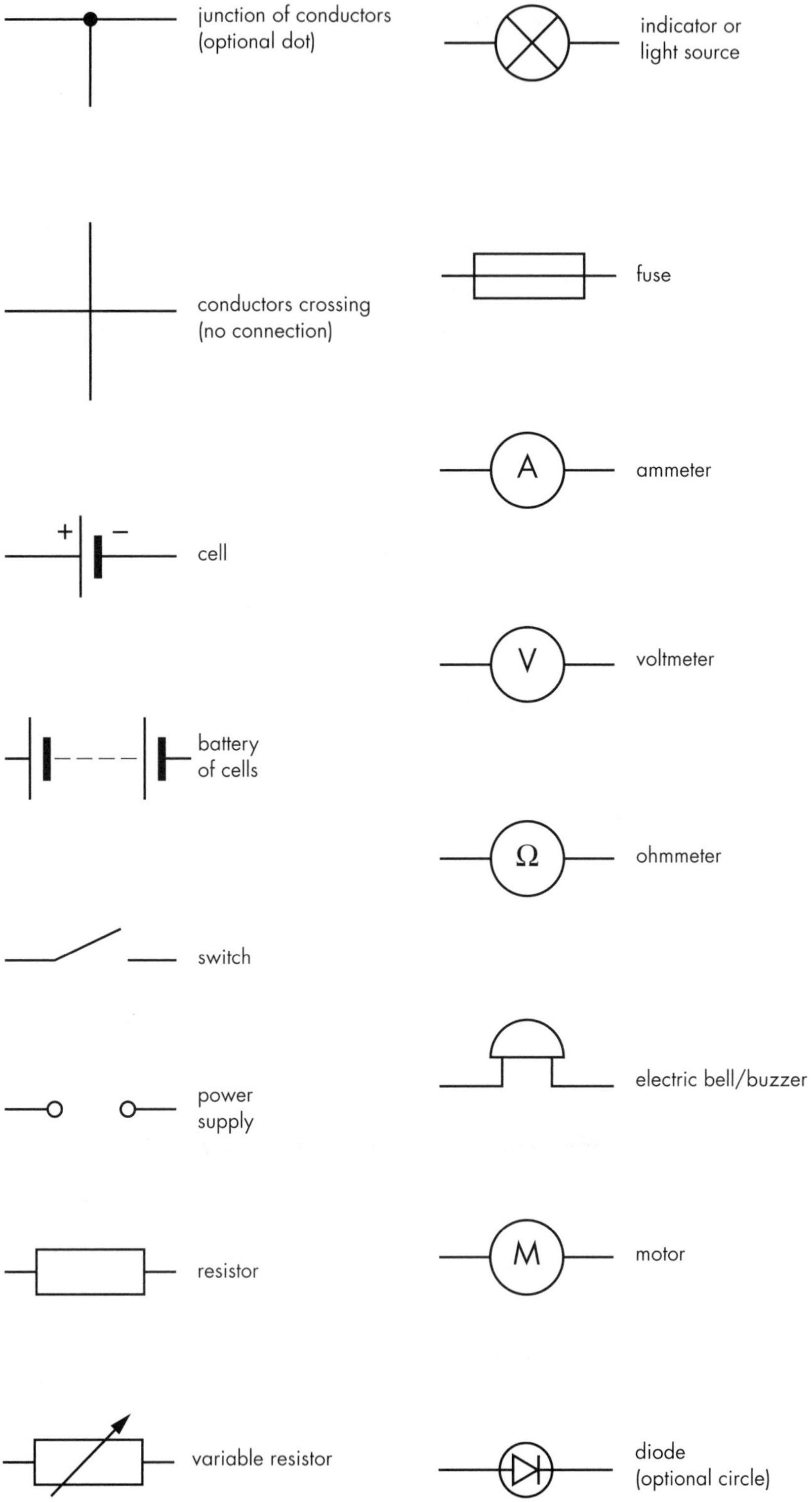

junction of conductors
(optional dot)

indicator or
light source

conductors crossing
(no connection)

fuse

ammeter

cell

voltmeter

battery
of cells

ohmmeter

switch

power
supply

electric bell/buzzer

resistor

motor

variable resistor

diode
(optional circle)

Introducing resistance

Start by revising conductors and insulators. Show pupils a
selection of materials (e.g. various metals, plastics, wood, etc.)
and ask them to predict which will conduct an electric current.
Using a simple circuit with a battery and bulb, check their
predictions. Point out that the glowing filament of the bulb
shows that an electric current is flowing through it. This will
help pupils to begin to refine their notion of 'electricity'.

You might also give pupils the opportunity to look closely at
a variety of filament light bulbs. Can they see the filament wire
inside the bulb, and the two connections where the current
flows in and out?

Is there anything in between conductors and insulators
which lets a little current through, but not much? Introducing
resistance by posing this question will make it easier for your
pupils to understand the effects of bulbs in series and parallel,
rather than just learning off by heart what happens.

Demonstrate varying resistance using a carbon rod or a
pencil lead, so that they have to look very closely to see that
the bulb is just glowing. You will need two crocodile clips to
make the circuit shown in Figure 4.3. Keep one still while you
move the other one along. It produces the same effect as a
dimmer switch.

Use the term 'resistance' and ask them to describe in their
own words what is happening, e.g. 'current just trickles
through', 'it has to fight to get through'.

Figure 4.3
*The resistance of
the carbon rod is
greater than that
of metal wire, so
that only a small
current gets
through. The
amount of
resistance
depends on the
length of the rod
in the circuit.*

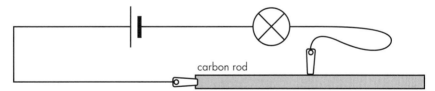

carbon rod

Modelling current in a circuit

Pupils accept the idea that a complete circuit is necessary for a
current to flow, but then they have to grasp the idea that
electric current is the same all the way round a circuit. You can
make a model (Figure 4.4, overleaf) which will form the basis
of a discussion. Use a length of the chain sold for basin plugs,
together with some plastic tubing and a cotton reel. Pass the
chain through the tubing, join the ends, and pass it round the
reel. Turn the reel (which represents the battery) and the chain
(current) passes through the tubing (wire). Pupils should be
able to identify the analogues of wire, current and battery.

Figure 4.4
A model of current flow in a circuit. (Adapted from: The World of Science, 1997, p.18. Association for Science Education/John Murray, London.)

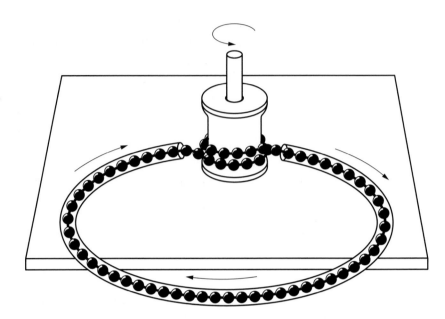

The advantage of this model is that the chain moves all round the circuit at the same speed. This stops pupils believing that the current is greater before the bulb than after it, which is very common!

You can even squeeze the tube (increasing resistance where your fingers squeeze the plastic!) and show that the current decreases. Unfortunately this model will not explain the working of a circuit with parallel paths for the current.

Some pupils, and many adults, believe that current flows (i.e. electrons move) very fast because of the speed at which the circuit responds and the bulbs light up when it is switched on. If they ask about this the best answer is that there are so many billions of electrons in a length of metal wire, each one of which is carrying a charge, that they only need to move at a few millimetres per second to make several amps of current. (This speed is called the 'drift velocity' of the electrons, but they do not need to know this – the greater the drift velocity, the greater the current.) In terms of the model shown in Figure 4.4, as soon as one segment of chain enters the tubing, another leaves the tubing at the other end.

Ammeters and amps

Current is a flow of electric charge. Teachers often talk about current in rather abstract terms, as 'something flowing' within wires. Pupils may find it easier to think more concretely of current as a flow of charged particles called electrons.

Try to get across the idea of how much current is flowing. Since this is related to the number of electrons passing a point every second, it will depend on how many electrons are moving and also how fast they are moving (Figure 4.5).

Figure 4.5
*Current as a flow
of electrons.*

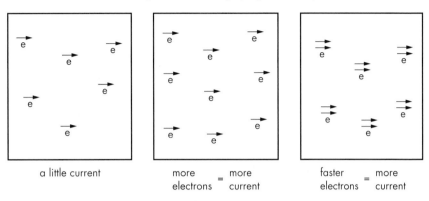

a little current

more = more
electrons   current

faster = more
electrons   current

This might be a good moment to try more than one battery using an ammeter in the circuit. They may know a little about this from their primary school work but are unlikely to have used an ammeter. The current in a circuit with one or two batteries (cells) and one or two torch bulbs would need an ammeter reading from 0 to 1.0 A. If using analogue (needle-and-scale) meters, pupils may have trouble with the decimal point, especially where the divisions on the ammeter scale are 0.02 A! If using digital meters, however, pupils may not see the patterns because they are confused by the availability of two decimal places. (For example, one battery may give a current of 0.08 A while two give 0.17 A, which is not obviously double.) A simple approach is to use an analogue meter but ignore the scale; mark the pointer positions on the meter.

Pupils should find that two batteries (in series) push twice as much current around a circuit. With two bulbs in series, the current is halved, so there must be twice as much resistance in the circuit. Problems may arise because apparently identical bulbs may be slightly different and one may look brighter than another even though the same current is flowing through both. Also, a bulb's resistance increases as it gets hotter, so two batteries may not make twice as much current flow. You may have to use a resistor in place of a bulb to show this.

In some ways it is easier to begin using a power pack instead of a battery. Because these power packs are plugged into the mains they look very dangerous to the pupils. It's a good idea to show that it is perfectly safe to touch bare wires attached to a 12 V power pack. Demonstrate this by showing, very slowly with much teasing, what happens when you hold the bare end of a wire connected to one terminal and then the other. Then hold one in one hand and the other in the other hand at the same time. (They think you are a hero!)

! *Make sure that pupils do not think that they can do this at home with the 230 V supply.*

Point out that 12 V is not enough to push a significant current through you, but the 230 V mains supply would. The idea is to ensure that they are not frightened by using low voltage. Some teachers stress danger so much that the more nervous pupils refuse to touch any part of a circuit, or indeed to think about electricity at all.

The lamp to use with a power pack is a car headlamp bulb which takes about 2.00 A when connected to a 12 V supply. (This is roughly ten times the current which is used to light a little torch pea-bulb.) Ones with the same current rating look more similar than torch bulbs because they are manufactured to a tighter specification.

The power pack should be set at 12 V d.c. – use the red and black terminals. There are also yellow (a.c.) terminals. Both will light a bulb, but a.c. will not register on an ammeter if you use one and may burn it out. It's a good idea to tell the pupils that power packs must always be set at 12 V d.c. for the time being.

There might be questions here about volts and amps but don't go into it properly. It's enough to tell them that volts show you how 'strong' the power source is – how much of a push it provides to make a current flow – while amps tell you the size of the current.

Just a couple of other points here. Always use a bulb (or other component with resistance) in the circuit with the power pack to stop the current being so great that it will damage the power pack, or the ammeter, or both. Most power packs have a 'trip' (a resettable fuse) to protect them from overload.

Secondly, a thin wire connected to a power pack could get rather hot. Avoid thin wires; always use insulated wires with plastic coating.

Tracing current
Current flows from positive to negative. (Electrons, being negatively charged, flow from negative to positive, but try to avoid this complication at this stage.) Ask pupils to trace the path of the current from the positive terminal of the power supply to the negative. They should say that it flows *through*

the ammeter, *through* the bulb and so on. They can also trace current paths on circuit diagrams. Later, this will help to emphasise the difference between current and voltage, since voltages are measured *across* components.

Resistance of thick and thin wires

Which has more resistance, thick wire or thin wire? Pupils can investigate this using simple equipment as shown in Figure 4.6. Use bare nichrome wire; copper has too low a resistance, and steel rusts. A suitable diameter is 0.37 mm (SWG 28), which has a resistance of about 10 Ω per metre. Pupils can twist two or more lengths together to vary the thickness.

Figure 4.6
Investigating resistance. A length of bare wire is taped to a metre rule. Including a bulb ensures that the current does not become too high.

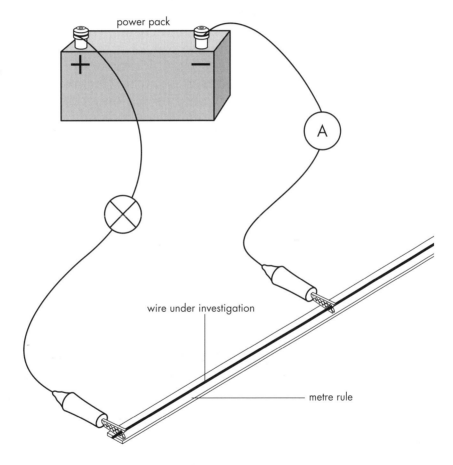

Pupils should decide how to make this a fair test. (Same voltage, length, type of wire, etc.) Some might repeat their measurements using 50 cm of bare wire. There is no need to cut the wire, just move the crocodile clip down it. They should make a line graph if they are able.

Series and parallel circuits

Some pupils may well have done this in primary school but it is most unlikely that they understand it. Keep getting them to imagine what is going on in terms of current flow.

Use one bulb and then two bulbs in series in a circuit. (See the *Equipment notes*, page 185.) Ask why the two bulbs are dimmer. Why does less current flow? (They usually say, 'Because the two bulbs have to share the current.') Get them to look at the filament of the bulb, perhaps using a magnifying glass. It is as fine as a hair. (Edison used a carbonised hair from a man's beard as the filament in his very first light bulb!) What do they know about the resistance of a very, very thin wire? So now can they explain why so very little current flows? Use the chain model to explain: there are two narrow points that the current has to squeeze past, so it is twice as hard for the current to flow.

Now put two bulbs in parallel (Figure 4.7a). It is a good idea to talk about how the current flows round the circuit and to demonstrate the circuit before asking them to build it. The question the pupils need to try to answer is: Why are the bulbs still bright and not dimmer?

Some pupils will answer that some current goes one way and some the other. They may even guess that exactly half the current goes through the top bulb and the other half goes through the bottom one.

The first thing to do is to check on this. They should be encouraged to move their ammeter round to different places in the circuit. They should conclude with something like: 'The currents separate and then join up again.'

The next thing to find out (if the pupils are doing well) is whether the two lots of current add up. You will probably have to demonstrate this yourself. Use the same ammeter in two different places and use a rheostat (a variable resistor) as well (Figure 4.7b). Now, as we say in our proverb, the current 'chooses the path of least resistance'. Strictly speaking, the path of lower resistance gets a bigger share of the current. Try to get the pupils to talk about it in their own words. With three ammeters in the circuit, it is a kind of adding machine!

Figure 4.7
a *Two bulbs in parallel. Add the ammeters later when you want to show how the current divides and then recombines.*
b *Adding a variable resistor allows you to make the two currents different.*

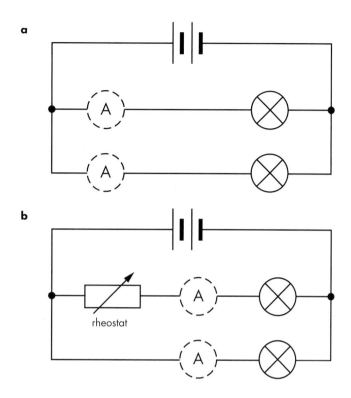

Now ask: Why does more *total* current flow when there are two lights in parallel? Everyone finds this difficult. Try explaining by using an analogy with traffic. If you want to go from place A to place B and the road is not very good the flow of traffic will be slow. However if a second route is opened, then there will be twice the flow of traffic. (They say half goes one way and half the other – that's true if the two paths are the same, i.e. they have the same resistance.) If one pathway is narrower then less traffic will flow along it than along the other one.

Later, when they have learnt about voltage, they may think about this in a different and more advanced way, but this is quite satisfactory for now.

◆ *Further activities*

♦ Extend the investigation of the resistance of nichrome wire into making a dimmer, with a loose wire sliding up and down along the fixed metre wire. This is very popular with the younger pupils.

4.2 Magnetism

♦ *Previous knowledge and experience*

Most pupils will know that magnets attract iron and steel. In primary schools, the emphasis is often on the 'poles' of the magnets. Now we want to focus their attention on the force between the two poles, and eventually on the region where this force acts – the 'field' of magnetic force.

♦ *A teaching sequence*

The capacity to pick up a whole chain of paper clips is often used to 'measure' the strength of a magnet. It will be useful in other experiments. This leads to a simple experiment to put in rank order a number of magnets by hanging a one-by-one chain of clips from each, until the last one drops off, and counting them.

Magnetic materials

Unfortunately many pupils use the words 'steel' and 'iron' for all metals so it is worthwhile pausing here to explain that there are many other metals, most of which are non-magnetic. There are two other metal elements that are magnetic, but weaker. These are nickel and cobalt, but the pupils are unlikely to come across samples of them. The magnetic properties of iron and steel are slightly different. Soft iron is easy to magnetise and demagnetise. Hard steel is difficult to magnetise, but retains its magnetism well. Note that some forms of stainless steel are non-magnetic.

Figure 4.8
A chain of paper clips shows how strong the magnet is; here, the force passes through a piece of cardboard.

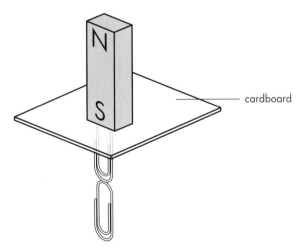

cardboard

The magnetic force of attraction for iron or steel seems to 'go through' several substances (Figure 4.8). Provide a tray of different substances: paper, cardboard, different textiles, wood of different thicknesses, plastic, aluminium foil, etc. Ask your pupils to find answers to these questions:

- What substances does the force of attraction of a magnet go through?
- Is the magnetic force any weaker on the other side?
- Does it make any difference how thick the substance is?

For the second and third investigations pupils will need a small store of paper clips, but they should ask for them themselves without prompting.

Attraction and repulsion

Unlike poles attract, like poles repel. Again some of this will not be new to most pupils, but it is probably necessary to go through it again.

They may say that a north and a south pole 'stick' together. It is important to clear up the difference between 'sticking' and 'attracting'. They should be able to feel the force between the magnets when they hold them a little way apart. Get them to explain what attraction means using the word 'force'. (It means that there is a force on the north pole, pulling it towards the south pole, and one on the south pole, pulling it towards the north pole.)

Repulsion seems much more weird than attraction. Once again they should be able to explain the meaning of repulsion using the word 'force'. (This time the force on one pole pushes it away from the other.)

Here are two extra activities; these will be useful when studying magnetisation later and for pupils who want to go faster because they learnt all this before in primary school:

♦ What is there in the middle of a bar magnet – a north or a south or neither? (A magnet is just steel in the middle, neither a north pole nor a south. Give the pupils a pair of magnets and they should be able to find this out for themselves without being told how. The middle of each magnet is attracted by both the north and south poles of the other magnet. It is repelled by neither.)

♦ Which of these two identical metal bars is a magnet and which is a piece of unmagnetised steel? (They will need to do 'ends to middle' to work this out. A magnetic pole will attract the middle of a bar or a magnet. A non-magnetic 'end' will not. Quite a teaser!)

Action at a distance

This is what fascinates the pupils, just as it fascinated Isaac Newton about gravity more than three centuries ago. Why do forces do this? The pupils ask simple but dreadfully difficult questions that cannot be given proper answers at this point, like 'What is happening in between the magnets?' and 'Why do two north poles repel each other?'. The best approach is to set them some investigations first, as below:

♦ How far apart can you feel the attraction or repulsion? Measure it! (A good way to start this is to stand up one magnet on its end and bring up the other magnet slowly until the first one topples over. Alternatively, have one hanging up in a fold of paper on a loop of cotton and approach it with one pole of the other one.)

Figure 4.9
a Two ring magnets repelling. Add Plasticine to increase the downward force. The force of repulsion between the magnets is equal to the weight of the upper magnet.
b Measuring the force of attraction between two magnets.

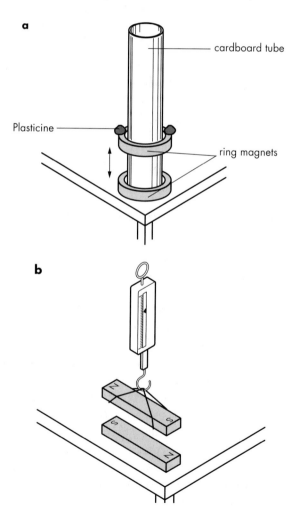

♦ Can the force of repulsion be measured? (Two ring magnets on a cylindrical cardboard tube are arranged so as to repel each other (Figure 4.9a). Why does the top one not fall? How far apart are they? Obviously the top one is being kept up by a force as big as its weight. Give them a newtonmeter and some Plasticine to increase the weight of the top ring. Can they measure the force of repulsion at different distances?)

♦ Can the force of attraction be measured? (Some pupils might have an idea how to do this. Try the arrangement shown in Figure 4.9b. With paper and wood to keep the magnets apart more measurements can be taken at closer distances, but the lower magnet must not come off the table.)

These are not accurate experiments but they do show how steeply the force falls away as the magnets move further apart. Quicker pupils might draw graphs of force against distance.

Magnetic fields

Now pupils can use iron filings to look at the shapes of magnetic fields. This should begin to make sense because they have just investigated forces at a distance. The magnetic field is the area around a magnet where this force operates. Of course it gets weaker further away, and it is really three-dimensional all round the magnet.

Pupils should avoid touching or blowing the iron filings as they are easily transferred to the eyes.

   Iron filings can be messy. It is a good idea to place a card on top of the magnet, and sprinkle the filings on the card. Then they can be readily returned to their pot. Better still, keep the filings confined within a Perspex box.

   If there are two magnets the situation becomes more complicated, with the force towards one pole combining with the force away from the other; the result is a pattern of curved lines of force.

   Pupils do not easily see the patterns of lines of force. Some just see splurges of iron filings. The lines show up better if you tap the card or box carefully. It is helpful to show them the conventional representations (Figure 4.10, overleaf) and ask them to match their filing patterns to these standard diagrams. Convention is to draw arrows from north poles to south poles.

Emphasise that the force is stronger nearer to the poles where the iron filings are pulled right on to the ends of the magnet. This is represented by the lines of force being closer together. There is neither north nor south in the middle of a magnet, so there are no lines of force there.

Figure 4.10
Conventional representations of magnetic fields.

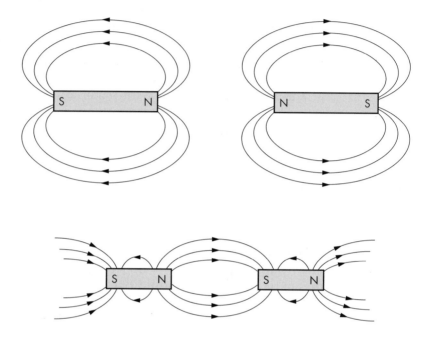

Using a plotting compass

Another way to draw the lines of force uses a compass. (The little iron filings are themselves like tiny magnets, and so the compass, which has a pivoted magnetic needle inside it, shows the lines of force in almost exactly the same way.) This method is more difficult than using filings because the pupils only have one compass so they have to record how the ends of the needle lie by making a pencil dot at each end, before moving the compass on. In the end there will be a trail of dots to be joined up to show the line of force. The pupils should start their plotting at one end of the magnet.

You could ask your pupils whether they think they can tell what the forces between the magnets are (attraction or repulsion) by just looking at pattern of the lines of force. Do they think the lines themselves pull or push?

Sometimes you can get quite good group discussions about these lines. Some pupils are quite good at verbalising how they see them. Michael Faraday, who was either the first to imagine these lines of force from patterns of iron filings, or the first to take them seriously, thought of them as being in tension – rather like rubber bands! Imagination is very important in physics.

Finally, you or your pupils might talk about gravity as being another field of force with lines. But the patterns they make are different. The Earth attracts us all towards its centre, but there is no gravitational repulsion. You might try to get your pupils to imagine what it would be like if we were repelled from the North pole and attracted to the South. (We might walk about leaning over towards the South!)

The Earth's field

The magnetic field of the Earth is due, we think, to currents in its molten iron core, so it will come up again in Section 4.5 on electromagnetism.

In discussion, the pupils may tell you that the Earth has a north magnetic pole at the North and a south magnetic pole at the South. It is not quite as simple as that; in fact, this is the opposite of the truth.

Ask your pupils: Given an unmarked magnet how would you find out which end of it was the north pole? Before they start, make sure everyone knows in which direction North is from the classroom.

Many of the pupils may know that you hang up the magnet to find geographical North. The best way to do this is to make a fold of paper, put the magnet inside it, suspend it by a loop of cotton (not from an iron clampstand) and wait until it stops swinging to and fro (Figure 4.11). The north magnetic pole will point to the North.

Figure 4.11

Suspending a bar magnet so that it is free to rotate.

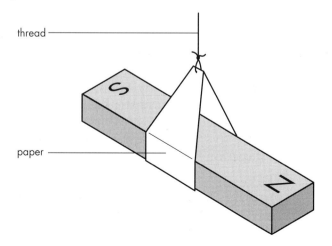

thread

paper

Sometimes the magnets have 'N' and 'S' marked on them and yet when hung up it is the 'S' which points to the north. Don't worry! This only means that the magnet has become re-magnetised in the opposite direction by a stronger magnet which happened to be stored near it.

The north poles of magnets are the ones that point to the North. They used to be called 'north-seeking' poles which makes it clearer. The smart pupil who deduces that the pole at the North (in the Arctic) must be a magnetic south pole is quite right. Some teachers prefer to call the poles 'north-seeking' or 'south-seeking' for the first term but pupils must get used to the usual terminology eventually.

Another complication is that geographical North is not the same as magnetic North. The geographical poles are where the Earth's axis of rotation emerges from the Earth. Magnetic North is at some distance from 'true North' and is slowly moving. No one is quite sure why it does this – a mystery! And occasionally the Earth's field completely reverses, so that one day compasses will point in the opposite direction!

Magnetising

For this work and that in *Modelling a magnet* opposite you will need a fairly large nail, about 3 inches long. The pupils will also need some iron wire which can be cut with a pair of pliers, tongs, a Bunsen burner, safety glasses, paper clips, a magnet, and a compass.

Figure 4.12
Magnetising an iron nail.

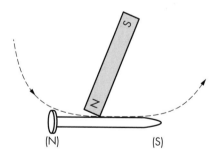

(N)   (S)

Most pupils will already know about making a magnet. They hold the nail down and stroke it firmly one way with one pole of the magnet (say the north pole) as shown in Figure 4.12. It is helpful to say that it is rather like combing out your hair in one direction only. Ask your pupils these questions:

- How can you prove that the nail is now magnetised? (The best way is using the paper clips, as before, but a compass will tell you too so long as only one end of the nail attracts the north of the compass and the other end repels it.)

- Which end of the nail has become magnetised north? (Now they do have to use their compasses to show that the north end of the compass needle is attracted to the last bit of the nail to be stroked by the north pole of the magnet.)

One problem that turns up fairly often is that the compass needle has been re-magnetised in the opposite direction. The best way to cope with this is to begin with everyone testing their compass to see which end of it points to the North. This is the north pole of the compass. Make sure that they have their magnet a good way off before they do this test.

Discuss with the pupils why stroking one way with a magnet magnetises the nail. The word 'combing' often suggests making hair all lie in one direction. That is not a bad start. Of course the tiny magnets, or magnetic regions, inside iron and steel objects are not rooted at one end like hair, but the pull of the magnet does align them all in the same direction.

Modelling a magnet

This model is imagined, rather than constructed like the model of electric current in Section 4.1. It is very valuable to get pupils to use their imagination in this way. It is a good 'physics' approach to explaining things. It is a beginners' approach to the theory of magnetic domains.

Figure 4.13
Using a microscopic model to explain **a** *magnetisation, and* **b** *why a magnet cut in half makes two magnets.*

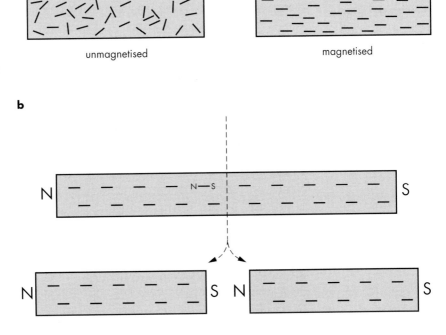

a

unmagnetised

magnetised

b

We imagine iron and steel to be full of tiny magnets, or magnetised regions (Figure 4.13). If they are all aligned in one direction there is a north-seeking pole at one end and a south-seeking pole at the other. The material is magnetised. If the tiny domains are disordered, clinging to each other in circles and other shapes, they cancel each other out and the total effect is an unmagnetised sample.

It is a good idea to get the pupils to draw this model of magnetised and unmagnetised steel. When they have understood the model, get them to carry out the following experiments and to explain them using the alignment model:

♦ Does dropping a magnetised nail on the floor several times reduce its magnetic power, and if so why? (They need to use the paper clips to measure the strength of the magnetised nail first. The easiest explanation is that falling on a hard surface knocks the little internal magnets out of position.)

♦ Does heating a magnetised nail red-hot in a Bunsen flame, then cooling it on a mat, reduce its magnetisation, and if so why?

! *Pupils must wear safety glasses and use tongs to hold the nail.*

(They may have to re-magnetise the nail before they begin. This experiment works best if they leave the nail to cool down in an East–West position on the mat so that the Earth's field does not affect it. The easiest explanation is that heat makes the little magnets all jog about and come out of alignment.)

♦ If a piece of wire is magnetised and then cut in half, does it become two magnets with new poles where it is cut, or two half magnets with only one pole each? (A 10 cm piece of steel wire will not be a very strong magnet so they have to test it with a compass rather than with a chain of paper clips. Those pupils who do this carefully find that the poles at either end are unaffected and there are two new poles, a north and a south, in the middle. This is because the little magnets are internally aligned inside all the way through. From the outside they have no effect. However, wherever the magnet is cut through there will be a north pole on one side and south pole on the other.)

Further activities

♦ Ask pupils to design and make a toy which uses magnetism.

♦ Pupils can make their own compasses (or use plotting compasses) and use them to find magnets hidden around the room. They could draw a map to show where they are hidden, and where their north and south poles are. What other magnetic materials can they find?

♦ Can they magnetise a piece of steel so that it has north poles at both ends, and south poles in the middle? How can they show that they have done this? (Stroke with a south pole from centre to one end, then to the other; use iron filings to show the pattern.)

Enhancement ideas

♦ High temperatures (above about 770°C) destroy the magnetism of iron and steel. The Earth's core is hotter than this, so there is no permanent magnetisation there. Hence we deduce that currents in the core are responsible for the Earth's field. The rare reversals of the Earth's field occur when these currents reverse. It is hard to imagine a permanent magnet inside the Earth reversing.

♦ We know about the Earth's field because of the existence of lodestone, a magnetic form of iron oxide, which was used for compasses for thousands of years. If people hadn't made this discovery, we might never have discovered electromagnetism, motors, generators, etc. How different life might have been!

♦ Christopher Columbus was unaware that magnetic North did not coincide with geographical North. This led him to travel further south than he had intended, with the result that he made landfall in the Caribbean. At the time, his crew were restless and anxious to return home. It is believed that if he had followed his intended path, he would have turned back before he ever reached land.

4.3 Electrical energy transfers and voltage

◆ *Previous knowledge and experience*

This section of work is slightly more difficult than the previous two and it is unlikely that your pupils will have learnt any of it in primary school. Although it builds on what has been taught about electric circuits in Section 4.1, it also uses ideas of energy and its transfer. You will need to co-ordinate this work with your teaching of energy as outlined in Chapter 1. An electric current is a way of *transferring* energy, rather than a form of energy. It is better to talk simply about 'energy', rather than 'electrical energy'.

◆ *A teaching sequence*

In order to avoid the common muddle between amps (current), volts, and watts (energy transferred per second), this work has been separated from the section on electric circuits where the pupils learnt about current and resistance.

Energy transferred electrically

Small electric motors are quite easy to get hold of from a model shop if there is not one in the laboratory. Connect a motor to the required voltage (power pack or batteries) and you should be able to make it turn round, pull up a weight or haul a trolley up a plank of wood. Ask the pupils where the energy to do this work comes from. Most will say that it was stored in the battery or supplied by the power pack. Reduce the voltage and the motor will work more slowly.

Another demonstration of the transfer of energy by a current is to make a coil of wire that will glow red-hot when a current flows through it. Cut a 70 cm length of nichrome wire, about SWG 26 (approximate diameter 0.5 mm), and coil it round a biro leaving straight lengths at either end. Remove the biro. Join the ends to the power pack and gradually turn up the voltage until the coil glows red-hot. It is quite spectacular! It is possible to show, visually, that the heat produced seems to depend on the number of volts (turn up the power pack) and also on the current (a shorter length will provide less resistance so more current will flow with the same voltage setting).

Towards a definition of volts

It is usually far easier to begin with simple exercises about volts, and only later to start defining potential difference which pupils find really very difficult. Most of them have already heard about the 230 volts of mains electricity, 1.5 V batteries and so on, so we begin by relating the volts to joules of energy.

The number of volts is equal to the amount of energy given out per second per amp of current:

$$\text{volts} = \frac{\text{joules}}{\text{amps} \times \text{seconds}}$$

To make this equation 'come alive' you will have to talk it over and use some examples:

1. Suppose we have a light bulb which can produce 60 joules every second when 2 A of current are flowing through it. How many volts will it be working from? (30 V)
2. Suppose we have another bulb which works on the 12 volt power pack, and has 0.5 A current flowing through it. How many joules of heat and light energy will it generate per second? (6 J/s)
3. An ordinary household light works on 230 V and also usually has about 0.5 A of current flowing through it. How many joules of heat and light energy will it generate per second? (115 J/s)

It is a good idea to talk over these last two examples. You might even show them the two light bulbs in operation (a small car bulb working off the 12 volt power pack and a 100 W household light bulb working on the mains). Their difference in brightness and heat are obvious. (Do not use an ammeter; the mains is a.c. and the chances are that you do not have an ammeter to measure it.)

This shows that it is not just the value of current in amps that affects the amount of energy that is transferred, but also the number of volts.

Connecting a voltmeter

If we want to measure the voltage of a battery, or the voltage across a lamp, we connect a voltmeter *across* it (i.e. in parallel). Your pupils will need to practise this: set up a circuit, *then* add the voltmeter. It will need two connecting wires, and you don't need to break into the circuit. Similarly, they should practise adding voltmeters to circuit diagrams.

Unlike current, volts don't flow. Always talk about the voltage *across* a component. This will make for an easier understanding of the term potential difference later.

Electrical power

To arrive at the total amount of energy given out per second by an electrical device, you need to multiply the number of amps by the volts. It is called the *power* and is measured in watts. Explain that more current flowing means more energy transferred, and a bigger voltage (bigger push) also means more energy transferred, so that it is evident that both current and voltage matter:

power = voltage × current $P = V \times I$

watts = volts × amps = joules of energy per second

Most pupils find it easiest to recall the equation in terms of units rather than the names of the quantities.

For homework the pupils can go round their houses looking at their electrical appliances and reading the label which gives the voltage (230 V) and the power (possibly in kW or kilowatts, that is thousands of watts). They should also look at battery-operated devices – a calculator may be labelled '3 V, 0.0004 W'.

A 2 kilowatt electric fire, for example, working at 230 volts, takes an amount of current they *should* be able to calculate easily!

Measuring energy transferred

What factors affect the rise in temperature produced in a beaker of water by an immersion heater? Pupils should realise that the voltage, the current, the time the heater is left on, and the amount of water are all relevant. Encourage them to ensure that very little of the heat generated can escape by using insulation and as little water as will just cover the immersion heater.

For this experiment each group of pupils will need both a voltmeter and an ammeter. The voltmeter is put across the points where energy is going to be transferred to the water (Figure 4.14). A 12 V immersion heater (the sort used for specific heat capacity measurements of solids) is ideal.

! *Pupils should wear safety glasses. If the heater has a damaged seal it could explode.*

Suggest a suitable volume of water so that pupils don't treat this as one of the variables.

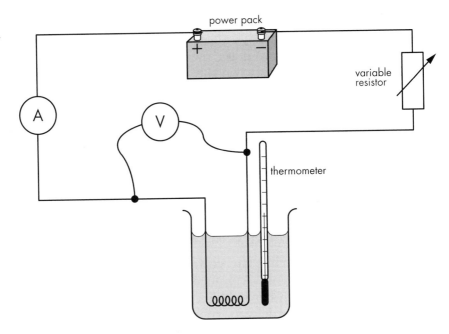

Figure 4.14
How does energy supplied depend on voltage and current?

They will soon find that if they change the voltage they will also change the current. Suggest that they multiply the two together and measure watts. The other variable is the time. Advise them to use minutes, or multiples of 100 seconds.

Modelling voltage in a circuit

Some people might want to put this work earlier in the course, but experience shows that the connection between volts and energy is the simplest approach. Now we can add that volts are a measure of the *potential difference* between two points in the circuit. Your pupils may already know about potential energy, so that helps us to move on. It is only 'potential' because if there is no connection between the two points, just a gap in the circuit, this potential difference still exists, even though no current will flow between the two points and no energy will be transferred. It is rather like two points on a hillside. Balls could roll downhill from one point to the lower one, gaining kinetic energy and releasing it as heat if they collided with a rock! But if the balls do not roll they still have potential energy.

If you have already introduced the model for a circuit shown in Figure 4.4, this is a good moment for getting it out again.

- The movement of the chain segments is like the flow of electrons, that is the current.
- The turning of the cotton reel provides the total energy which makes the current flow.

This energy is transferred to the segments by the tension in the chain. You can show, especially if someone helps by squeezing the tube to make more resistance, that there is a difference of tension between the chain where it is being pulled into the cotton reel, and where it loops back. This difference in tension in the chain is what gives the segments enough energy to move, and corresponds to potential difference (voltage).

Not all the pupils may follow this, but most will. Some teachers prefer to use a model of water in tubes, such as a central heating system or water flowing through a hose pipe when you put your foot on the pipe. You will need to stress that no current can flow unless there is a potential difference.

Resistance and Ohm's law

Measuring the voltages across different resistances
Each group will need a power pack, a voltmeter supplied with wires and crocodile clips, an ammeter, and a set of resistances – different lengths of resistance wire and/or lamps. It is wise to get your pupils to set up the circuit with all the resistances in series with the ammeter first, before adding the voltmeter.

Then they use the voltmeter to probe across each of the resistances in turn. They need to record the readings and decide which one has the most resistance. They may say that the biggest resistance needs the biggest potential difference across it to make the same current flow.

The most valuable part of this experiment is this discussion where the pupils use their own words to describe and explain why there is more potential difference across the largest resistance when the same current is flowing through all of them.

Defining resistance
Now you can go on and get them to put together (with your help) a qualitative version of Ohm's law:

- the more the resistance, the less the current flowing through it (for a given voltage);
- the more the resistance, the greater the voltage needed to make a particular current flow.

Pupils should thus see that, to work out the resistance R of a component in a circuit, we need to measure two things: the current I flowing through it, and the potential difference V across it, pushing the current through.

$$R = \frac{V}{I} \qquad \text{ohms} = \frac{\text{volts}}{\text{amps}} \qquad \Omega = \frac{V}{A}$$

Resistance tells you how many volts are needed to make an amp of current flow. This relationship is often referred to as Ohm's law, which says that, for many components, the current that flows through them is proportional to the voltage across them, i.e. double the voltage makes double the current flow. This is true for components such as resistors (provided they don't get hot), but not for filament lamps, whose resistance increases markedly as they get hotter. This is usually represented by current–voltage graphs (Figure 4.15). Note the gradient at any point is $1/R$.

Figure 4.15
a An ohmic conductor gives a straight-line I–V graph.
b A filament lamp is non-ohmic; as it gets hotter, it becomes increasingly difficult to push more current through it.

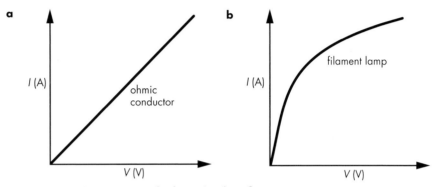

Human resistance and electrical safety

We are protected from electric shocks by the resistance of our skin. Our insides are good conductors, but dry skin has a high resistance, of the order of several thousand ohms. Wet skin conducts better – hence the saying that 'electricity and water don't mix'. The unit *230 V can kill* in *World of Science* (ASE, 1997) is useful for getting these ideas across.

◆ *Further activities*

- ◆ Look at how the resistance of thermistors changes with temperature. A thermistor (thermal resistor) has a resistance that changes (increases or decreases) rapidly over a narrow temperature range.
- ◆ Use an ohmmeter to obtain direct measurement of resistance.

◆ *Enhancement ideas*

- ◆ We are used to having ammeters and voltmeters readily available to us. This wasn't always the case. Alessandro Volta used batteries of cells to charge objects; he judged the voltage by touching them. The further the shock went up his arm, the greater the voltage!
- ◆ The symbol *I* for current was originally chosen as an abbreviation for 'intensity'.

4.4 Static electricity and charge

Electrostatics is considered either nothing but fun and sticking balloons on your tummy, or extremely difficult to understand. The way of teaching it suggested here uses the ideas already developed in the sections on electric circuits and electrical energy transfers, which has the advantage of revising all that, and expanding it, as you go along.

◆ *Previous knowledge and experience*

Pupils will be familiar with some electrostatic effects, such as what happens when you rub a balloon on your jumper, and the build-up of static when you brush clean, dry hair. You can incorporate these experiences into your teaching, and teach pupils to explain them.

◆ *A teaching sequence*

Charging by rubbing

For this you need a set of nylon rods and old woollen socks. These need to be clean and dry – put them in a very low oven for a while beforehand. You also need a set of polythene rods and nylon or silk cloths, again clean, dry and warm. Don't let the different cloths get muddled up or the residual charge on one lot may transfer to the other and give you anomalous results. Keep them in separate boxes. When rubbed, polythene rods gain a *negative* charge; nylon becomes *positive*. Do not mention positive and negative to pupils, however, until they arise in the course of these experiments. A suggested sequence of demonstrations is as follows.

- Both sets of rods and cloths, when rubbed together, give the rod a charge so that it can be seen to pick up fluff or small scraps of paper.
- Two similarly charged rods, when hung up in a paper fold, will repel each other (Figure 4.16). This is a better demonstration than charged balloons but the idea is the same. The effect gets weaker with distance.
- Two differently charged rods, e.g. rubbed nylon and polythene, will attract each other. (So there are two types of charge.)
- You can use this knowledge to identify which sort of charge is on a rubbed Perspex rod (or a biro), given that a polythene rod gains negative charge when rubbed. (To check the sign of an electric charge, try using a digital coulombmeter.)

Figure 4.16
Use paper and thread to suspend charged rods so that they can turn freely.

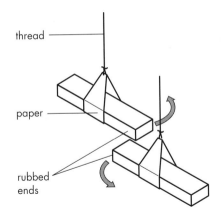

Conductors and insulators

You need a demonstration leaf electroscope, possibly with a lamp behind it, on your side, to throw a shadow of the leaf on the frosted glass on the other side. The shadow is much easier for the class to see than the leaf.

You need a rubbed polythene rod to give a charge to the electroscope. You could touch the top of the electroscope to transfer the charge, but this is unreliable. The polythene is an insulator so the charge does not move off it very easily. A better way to charge up the electroscope is to bring the charged rod near, then touch the top of the electroscope briefly with the other hand, and then take the rod away (Figure 4.17).

Figure 4.17
Charging an electroscope by induction. The leaf goes down when you touch it because electrons are repelled by the rod and run to earth through you. Take the rod away and the leaf goes up again because it lacks electrons. The electroscope is left positively charged.

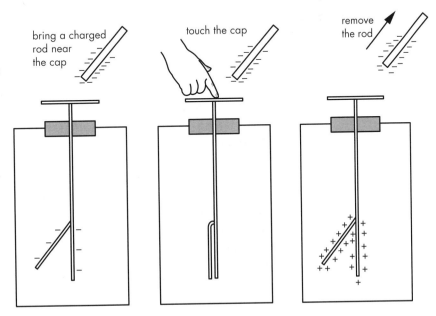

Now that you have a charged electroscope you can show the difference between conductors and insulators.

Touch the top with an uncharged plastic object. There is no effect – the plastic is an insulator. Touch the top with a metal object. A current flows through the object to neutralise the charged electroscope.

Charge up the electroscope again and touch the top with a piece of paper (dry or damp). Ask the class what would happen if one of them touched it with a finger, and let one of them do it. Even people can be conductors, albeit poor ones.

Beginning to model charge

Now you can begin to talk about charge in terms of *electrons*, since these are the mobile charged particles which are involved in these experiments.

- A charged object has extra electrons, or a shortage of electrons.
- All electrons have the same negative charge. Therefore they all repel each other. If there is a conductor there they will move away from each other along it.
- An uncharged rod has the same number of negative particles (electrons) and positive particles (which you can refer to as atoms or, more correctly, ions).

Recap some of the earlier experiments in terms of the movement of electrons, which can be readily rubbed from one material to another.

Can pupils work out why a rubber balloon, after being rubbed on your jersey, tends to stick to it, or why clean brushed hair is attracted to the hair brush?

Finally you could go back to metals and ask why a metal rod, held in the hand, does not seem to get a charge when rubbed. Of course they all know that metals are conductors and now need to link this up with the idea of moving electrons again.

Conservation of charge

You need an electroscope with a tin can on top of it for this. Take a nylon rod rubbed with a woollen sock, wrap the sock loosely around the rod and place both in the tin can (Figure 4.18).

Without touching the can, rub the rod up and down in the sock. There is no movement of the leaf, so no charge is being gained or lost. Now pull the rod out of the can, again without touching it. The leaf will go up – the sock has charge on it!

Figure 4.18
*Showing equal
and opposite
charges.*

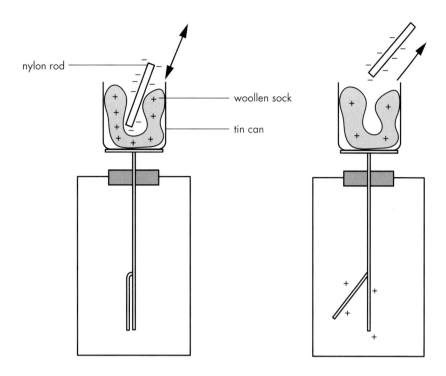

Now lower the rod back into the can. The leaf goes down –
the rod has the opposite charge on it. What's more, the
amounts of positive and negative charge must be equal, since
they exactly cancel out.

An alternative to an electroscope is a coulombmeter. With a
digital coulombmeter, pupils will need help in understanding
the changing values displayed.

Fields and current

All pupils seem to enjoy tricks with the Van de Graaff generator.
Some teachers do demonstrations involving charging up a pupil;
do not inflict this on any reluctant children, and only charge up
one pupil at a time to avoid large charges.

A Van de Graaff generator may be motor-driven or
hand-driven. Before you try out a Van de Graaff always wipe
the dome and the discharging sphere with a cloth dampened
with ethanol. Then dry them, rubbing with a clean dry cloth.
Make sure that there are no pieces of fluff left on them. You do
need to practise the demonstrations because they are spectacular
and most enjoyable when they work well. If it is a very damp day
you would do better to postpone this lesson to another day.

The following demonstrations do not take long and some are
very good for revision.

Using the air as a conductor

Charge up (switch on or turn the handle) with the dome and the discharging sphere about 1 cm apart in a darkened room. If you have cleaned them up as suggested above, you should get a spark crossing from the dome to the small sphere. Then, using the insulated handle, move them a little further apart. Again you should get sparking, but a little slower and bigger because it takes a very strong electrostatic field to break down the insulation of the air so that it will conduct a spark. The pupils will want you to go on and on but there is a limit. Take care not to point at the charged dome because your finger may act as a lightning conductor and you will get a bit of a shock. You can discharge the dome at any time by touching it with the earthed discharging sphere, again using the insulated handle.

Exploring the field of electrostatic force

When the dome is charged there will be a field around it with lines of force, rather like there might be round a single magnetic pole, if that were possible! The accessories for the Van de Graaff may include a bunch of threads: if not, it is easy enough to make what you want with a single plug to fit into the top of the dome and about twenty 25 cm lengths of white terylene thread (as used for sewing). When the dome is charged the threads stand up (Figure 4.19a). (If they do not they may be too long.) Clearly they are repelling each other, but, like iron filings for a magnet, they are also indicating the lines of force. Since charge tends to leak away from the ends of the threads you will need to keep up the charging process.

If you bring your hand, or a metal object, near the charged dome the threads will show you the change in the direction of the lines of force (Figure 4.19b).

Figure 4.19
a The threads spread out to show the shape of the electric field.
b Your hand alters the field.

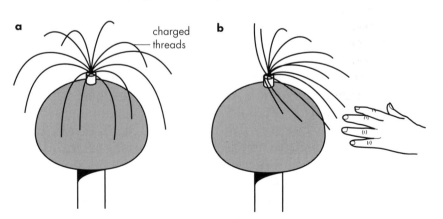

Showing that moving electrons are a current

Other accessories that should come with the Van de Graaff include a Perspex container with small polystyrene balls. This can be fixed on to the top of the dome and, when charged, the balls jump up and down for a bit and then stay up (Figure 4.20). The earthed small sphere needs to be some distance away from the dome. The balls have been charged and are repelled from the dome. Now if you touch the top of the container (not the dome) the balls will jump down, having given their charge away to run down your arm. The pupils can see this as a flow of charge carried by the balls, which goes down your arm as a current (but too minute to feel).

Figure 4.20

The movement of charged balls represents a current.

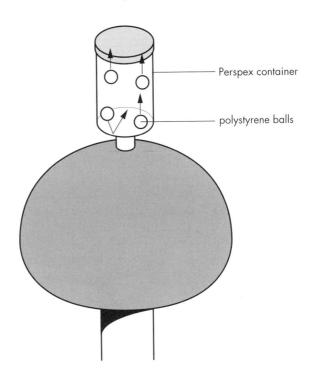

Perspex container

polystyrene balls

Another demonstration uses a neon bulb. Work in a darkened room. Stick one wire of the bulb down the socket in the top of the dome. Now charge up the dome. With the earthed sphere nearby, the bulb will light up every time there is a spark. If you move the small dome further away and touch the wire from the bulb with a screwdriver (holding the plastic handle), the bulb will light up if the circuit to earth is complete (Figure 4.21, overleaf). The sight of a bulb lighting up is a more convincing demonstration that current is flowing than anything else!

Figure 4.21
*When the neon
bulb lights up,
there is clearly a
current flowing.*

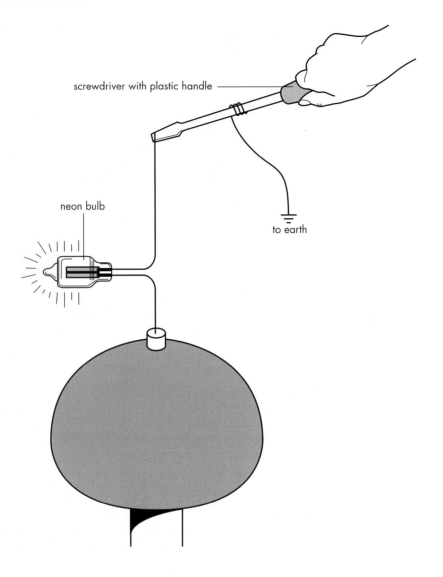

screwdriver with plastic handle

neon bulb

to earth

♦ *Enhancement ideas*

- ♦ The word *electricity* comes from the Greek word *electron*, meaning amber. Early experimenters rubbed amber as a reliable way of generating static electricity.
- ♦ It is unfortunate that conventional electric current flows in the opposite way to electrons in a circuit. This problem arose from the definition of positive and negative charges, which was formulated long before anyone knew of electrons.
- ♦ A current *can* be a flow of positive charges – for example, positive ions moving during electrolysis, or the current that flows through a neon bulb when it is lit.

4.5 Electromagnetism

♦ *A teaching sequence*

Electromagnetism seems difficult, so we start with what behaves like an ordinary magnet, in the sense that it has a north pole at one end and a south pole at the other. Of course its magnetism depends on the current, in ways that the pupils can easily discover for themselves. Only later do we go on to effects at a further remove from the simple magnetism we taught them at the beginning of this chapter.

A simple electromagnet

You could start with a demonstration where a strong electromagnet picks up a heavy load and you switch off the current so that it falls with a bang! You need a large, powerful electromagnet for this, capable of taking a current of 1 A or more.

> ! *Use a cushion or similar to protect the floor, and keep feet away.*

Then your pupils can carry out their own more careful experiments. You might want to use a Westminster Electromagnetic Kit and wind the electromagnets on C-cores, if these are available. Alternatively, you might prefer a more 'ordinary' set-up, winding wire around an iron nail.

Have a sample electromagnet set up, and show how it can attract a piece of steel (Figure 4.22, overleaf). Ask your pupils how they could measure the strength of the electromagnet (with a chain of paper clips or pins), and what factors might affect how strong the electromagnet is. This second question usually gets at least three answers:

- the amount of current flowing;
- the number of turns of wire;
- how long the current is kept on.

The first is right, the second is nearly right (actually it is the number of coils per unit length, so they should be prepared to wind one layer on top of another). The third turns out to be wrong. It suggests that they have a 'cooking' view of how an electromagnet works! It may be useful if one group actually shows that the strength of the electromagnet does *not* increase with how long it is 'on' for. (Do not turn it off while taking measurements.)

Figure 4.22
A simple electromagnet.

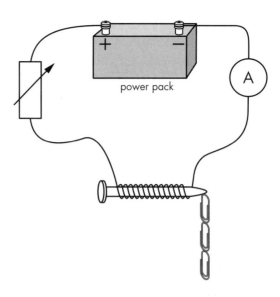

Occasionally pupils also suggest that it depends on whether the core is made of iron or steel. True, but that does not make a continuous variable and it is better demonstrated later.

With so many variables this has become a classic investigation. To get good results you need to have at least ten turns of wire, and you should increase the current (or the number of turns) in uniform steps, since once the nail is magnetised it will tend to stay that way, even if you reduce the current.

Your pupils will need about a metre of insulated copper wire, cotton or plastic-covered (commonly known as 'green wire'). Include a rheostat in the circuit, otherwise the circuit breaker on the power pack will switch off. Pupils will have to check their set-up with you before they switch on. Then they can gradually increase the current up to settings where the electromagnet just begins to pick up paper clips.

Magnetic poles and current direction

Your pupils can easily show that one end of their electromagnet is north and the other south using a compass. They can also show that sending the current the other way round the coils reverses the poles. Ask them: At the north end of the electromagnet, is the current flowing round clockwise or anti-clockwise? This is not as easy as it sounds. The pupils need to know that the current flows, by convention, from the positive terminal of the power pack round the circuit to the negative terminal.

Figure 4.23 shows you how they find out the answer and also a simple visual mnemonic for remembering it.

Figure 4.23
The poles of an electromagnet.

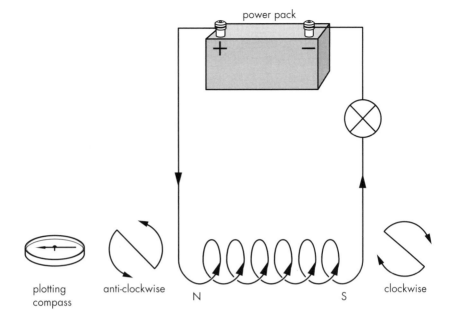

plotting compass   anti-clockwise   N   S   clockwise

Using electromagnets

The simplest case is when electromagnets are used, hung from a crane, to sort out iron and steel from scrap metal. Pupils will appreciate that you can turn the magnetism on and off; they may also be able to tell you that it is an iron core which is required, not steel, if the magnetism is to be reduced to a minimum when the current is switched off.

Other examples include a relay, where a large current can be switched on and off by a small one; and a circuit breaker, where any difference between the currents in the live and neutral wires causes an electromagnetic switch to open. (Note, however, that the trip in some power packs is not an electromagnetic device; it works using a bimetallic strip which bends due to the heating effect of an excessive current, thereby operating a switch.)

Magnetism without iron

This is where the topic becomes more mysterious. The pupils are not terribly surprised that the current flowing in coils round an iron core makes a magnet, but few expect the coils to be a magnet without the iron core. So let them find this out. The effect will be small but if they have a couple of amps flowing and at least 20 turns of wire the effect on a compass of turning off and on the power pack is quite enough to convince them.

Why does current produce a magnetic field? This is a hard question to answer and it is possible to tackle it in two ways.

First, explain that electricity and magnetism are really one effect, not two. Wherever there is magnetism there is always current of some sort; sometimes this is a current of electrons moving around inside the atoms of the iron or steel. Alternatively, you could suggest that a magnetic field is like the 'wake' of eddies that a boat leaves in water, so that wherever there is electricity, there is a magnetic field.

Show them that even a long straight wire carrying a current produces a magnetic effect (Figure 4.24). Use as large a current as your power pack will allow and gently nudge a compass round the wire when it is threaded through a piece of cardboard (possibly running down between two tables).

Figure 4.24
Demonstrating the magnetic field around a current.

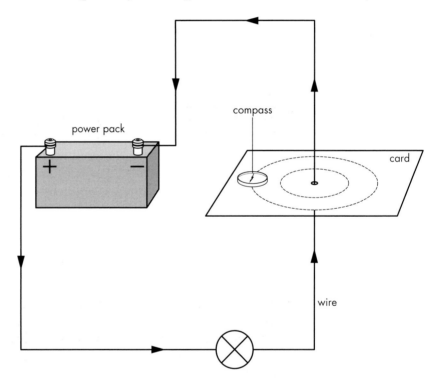

It is a good idea to place several compasses around the wire and then switch the current on and off. With the current off, they all point north; with the current on, they roughly form a circle round the wire, showing the shape of the magnetic field. The unexpected outcome is that the wire makes a magnetic field whose lines of force go round the wire. They do not come from one point and finish at another. This is a magnetic field without any poles!

The motor effect

If there are no magnetic poles associated with a current flowing down a wire, will it be attracted to a north pole, or repelled by it, or neither? The answer is that the wire does experience a force but it is at right angles to both the field and the current.

Your pupils can explore this effect for themselves using a loose and really thin strip of aluminium foil instead of a wire, as shown in Figure 4.25. When it is set up with just 1 amp flowing, they bring the north pole of a magnet near it, then a south pole, and then both, from opposite sides, at the same time. They may need some help to see that what happens is not repulsion, but that the foil is pushed to bulge out in front of the magnetic field or to retreat behind it.

Figure 4.25
Showing the force on a current-carrying conductor in a magnetic field.

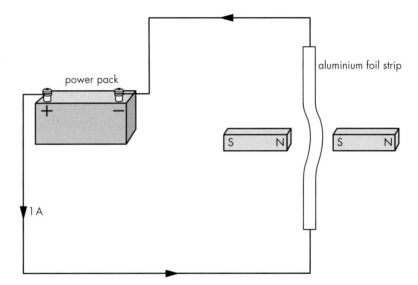

If you repeat this as a demonstration for the class, try changing over from d.c. to a.c. (remember to take out any ammeter first, or to change its setting to a.c.). The alternating direction of current flow makes the foil strip vibrate to and fro 50 times per second (the frequency of change of the a.c. mains). You can make this quite spectacular by moving the magnets nearer. (Try using two magnadur magnets on a yoke.) This time it is very easy to see that the direction of movement is at right angles to the direction of the magnetic lines of force between the north and south poles.

This is the moment to teach them Fleming's left hand rule if you want to do so. Try to avoid putting yourself in the position of showing the current flowing upwards. It looks very vulgar and you get the class falling about for ages!

Motors

Motors are very useful all over the home (kitchen, workshop and garden) and in industry. Even electronic devices like the computer, which work in quite a different way, often need a fan to keep them cool and this is little more than a motor with blades on it.

Most motors work by using the force on two sides of a rectangular coil in a magnetic field. The lines of magnetic force all run across the space where the coil is and are parallel to each other. There are two ways to explain how a motor works:

- You can regard the coil as an electromagnet. One side is a north pole, the other is a south pole. The coil is attracted by the permanent magnets, causing it to rotate.
- Alternatively, you can consider the forces on the current flowing round the coil. The force is *forward* on one side and *backward* on the other, so the coil turns.

Now, the clever thing is that the current reverses at the critical moment, so that the coil carries on round. The current may reverse because you are using alternating current, or because the motor is fitted with a commutator which reverses the connections twice during each cycle.

If you have a Westminster Electromagnetic Kit you can first demonstrate how a motor works and then let your pupils try for themselves. Follow the instructions and take special care over the starting position (Figure 4.26a and 4.26b). Start with the plane of the coil horizontal so that the lines of magnetic force are at right angles to the two sides of the coil; then one will go up and the other down.

Figure 4.26a
A model electric motor.
Your model should be given a preliminary check by holding the wires from the power pack against the commutator to make contact. (Adapted from: Physics Activities for GCSE, 1996, p.90.)

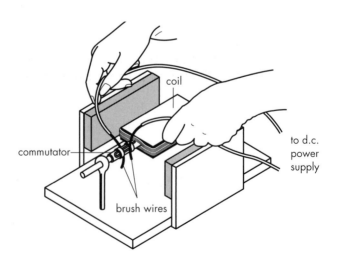

Figure 4.26b
*The brush wires can then be fixed so that you don't need to hold them.
(Adapted from: Physics Activities for GCSE, 1996, p.90.)*

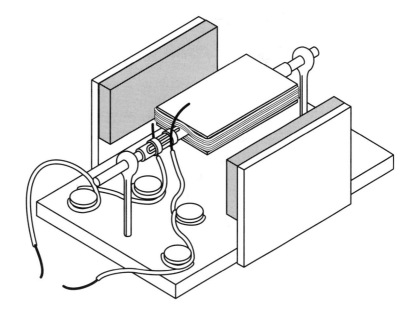

It's well worth getting this model to work, because pupils get a lot of satisfaction out of it. Try it yourself in advance until you are confident. Points to check if a motor won't turn.

- Check that the coil can spin freely inside the magnets and yoke.
- Check that the magnets have attracting faces towards each other.
- Is the coil horizontal at the start?
- Give the coil a nudge to start it off.
- Are the brushes bare wires?
- Are the brushes pressing firmly enough on the commutators?
- Is there insulation under the commutators?

Ammeters

Moving-coil meters are similar to motors, but they have a spring which resists the turning of the coil so that it only goes round through a large angle when the current is large. With a smaller current the coil turns through a smaller angle.

◆ *Further activities*

- ◆ Pupils can make model buzzers, relays and loudspeakers. You will find instructions in many published schemes. e.g. *Physics Activities for GCSE* (Whitehouse, 1996).

4.6 Electromagnetic induction

♦ *Previous knowledge and experience*

In their studies of energy resources, pupils will have studied some of the ways in which electricity is generated (see Chapter 1). Here, they will learn more about how generators work. You may have to explain the distinction between turbines and generators: a turbine is made to turn by moving air, steam or water (like the sails of a windmill); the generator is turned by the turbine and generates electricity.

♦ *A teaching sequence*

Making electricity

Electricity and magnetism go together. By this stage, your pupils have probably heard of 'electromagnetic waves'. Michael Faraday was the first to think about magnetic lines of force and made the first electric motor, and it occurred to him that it should be possible to make electricity flow by using magnetism. He tried and failed for almost 11 years. Then he hit on the idea of changing the magnetic field through another circuit with a switch.

There is a demonstration you can do, which is along the lines of Faraday's but simpler, to show this. You could try saying the following while you do it: 'Well, if when a current flows through a coil it makes the coil an electromagnet, perhaps we can make a current flow by putting a magnet into a coil. Who agrees with that?'

For this you will need a demonstration galvanometer (as sensitive as possible, reading in milli- or microamps), big enough for them all to see, and a coil with about 20 turns of wire (Figure 4.27). Put the magnet into the coil before you connect the galvanometer. Nothing happens when you connect the galvanometer into the circuit. Now pull the magnet out and, if they are watching (and if you do it quickly) there will be a sudden surge of current. This is electromagnetic induction.

Figure 4.27
A simple demonstration of electromagnetic induction.

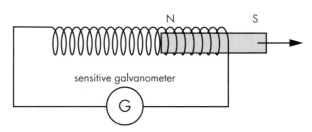

sensitive galvanometer

Your pupils can do the rest of the investigation for themselves. They will need one or two magnets, a fairly sensitive centre-reading ammeter, and wire to make their own coils. Here are two questions for them to explore:

- What factors affect the size of the current generated? (Speed of movement of the magnet, number of turns in the coil, strength of the magnet.)

- What factors affect the direction of the current generated? (The direction in which the magnet moves, which pole is approaching the coil.)

You should encourage pupils to picture the lines of force spreading out from the magnet's poles. As the lines of force cut across the wires of the coil, they induce a current in it. If you have a very sensitive galvanometer of the light-beam type, connect a single wire between its terminals. Move a magnet past the wire and you will see a flicker on the meter. This is the same effect as with the coil, but a coil is useful because the effect is multiplied up when you use many turns of wire.

Dynamos

Pupils may be familiar with dynamos used for lighting bicycle lamps. Dynamos work by electromagnetic induction. A first understanding is to think of a dynamo as a motor working in reverse. Connect a simple d.c. motor to a sensitive ammeter. Spin the axle of the motor and it will generate a small current. Pupils can picture the coil inside, turning in the magnetic field. As it turns, its wires cut through the magnetic field lines. (Note that bicycle dynamos work the opposite way – a magnet spins inside a coil – but the result is the same.)

   A demonstration is useful here (Figure 4.28, overleaf). For this you will need two pupils, each holding a magnadur magnet, and a large flat coil with a radius of about 15 cm and about 50 turns of wire. Connect the coil via long, loose wires to a centre-reading galvanometer. Turn the coil in the space between the magnets. You should be able to manage two turns of 360°, at least. As the coil turns the current flows first one way and then the other. The class should immediately see that the current keeps changing direction. It is a.c. – alternating current – which we use all the time at home.

Figure 4.28
*Demonstrating
how alternating
current is
generated.*

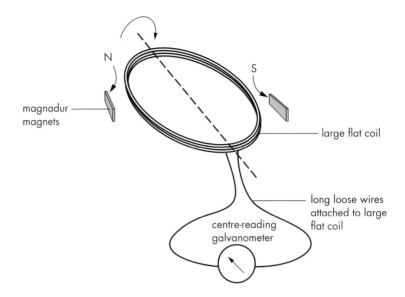

magnadur magnets

N

S

large flat coil

long loose wires
attached to large
flat coil

centre-reading
galvanometer

Terminology
The word *generator* is rather loosely defined, and could include
cells, photocells, etc. A *dynamo* is a generator that uses
electromagnetic induction. An *alternator* is a dynamo that
produces alternating current. Typically, alternators in power
stations spin 50 times per second to generate the mains
electricity we use every day.

Transformers
One of the big advantages of a.c. is that, by using a
transformer, it is easy to change the voltage and the current
without losing much energy. Transformers are all around us. A
child playing with an electric train might need a transformer to
get a smaller and safer voltage than the mains. So a 'step-down'
transformer is used. Many electrical appliances such as radios
and telephones use low voltages, transformed down from the
mains voltage. Electricity sub-stations serving local houses
transform the high voltage of the electricity grid down to
230 V.

At power stations, the generated voltage is transformed up,
perhaps to 400 kV, using a 'step-up' transformer. As the
voltage is stepped up, the current is stepped down, since the
power (= current × voltage) is constant. It is much better to
transfer energy across the country from the power station, via
pylons, to the users at low currents so that less energy is
liberated at heat. We don't want red-hot wires, and higher
voltages are all right so long as no one touches them while
standing on the Earth!

A transformer has an iron core with two quite separate coils of wire wound on it. Since electricity goes in to one coil, and comes out of the other, pupils immediately assume that they are connected together. They are – but magnetically, not electrically. If an alternating current flows in one of these coils (the primary circuit) it produces an alternating magnetic field which passes through the other coil (the secondary circuit). No current flows from one coil to the other; it is the changing magnetic field which induces the voltage. The voltage is stepped up or down in the ratio of the number of coils in the two circuits – twice as many turns in the secondary coil means twice the voltage out, but only half the current.

Nothing appears to happen when you use a transformer in the normal way except a quiet buzz. The whole point of it can be quite lost on the pupils, even if it is a 'demountable' transformer which you have put together in front of them. It just seems like a piece of rather dull equipment with two knobs here and two knobs there! You need to convince them that something interesting is going on, and a demountable transformer is the best way to do this (Figure 4.29a, overleaf).

> If the demountable transformer does not have proper mains connections to the primary coil, the supply must be isolated from earth with a hidden transformer to protect you.

Connect the primary coil to the 230 V mains. Then read the voltage across the secondary coil. You should have one, provided with the transformer, that gives a voltage of 12 V. Then put a 12 V lamp in this circuit. When you switch on the mains in the primary circuit the lamp in the secondary circuit will light up. Point out that the 230 V mains would have blown the bulb.

Now try to remove the top bar of the transformer. It is very difficult, or impossible, depending on the model. Turn off the mains and try again. The top comes off easily. This shows that, when the transformer is working, there is a very strong magnetic field in the core.

While the top is off, switch on again and you will find that the lamp is either not lighting up, or only just. Now bring the bar near. It is a good idea to put one end of it in contact with the top of the transformer and slowly swing it round until the other end touches the top. As more and more magnetic lines of force go through the secondary circuit the bulb becomes brighter and brighter. The vibration that you feel is due to the alternating current producing an alternating magnetic field.

Figure 4.29
a Stepping down the mains voltage using a demountable transformer.
b Lighting a torch bulb with a home-made coil.

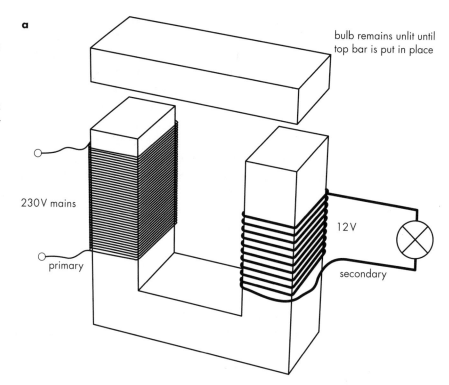

a

230V mains

primary

bulb remains unlit until top bar is put in place

12V

secondary

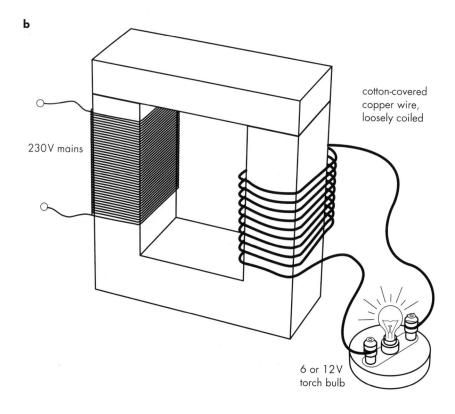

b

230V mains

cotton-covered copper wire, loosely coiled

6 or 12V torch bulb

A demonstration which really does surprise the class is a home-made circuit made from about 20 turns of cotton-covered copper wire and containing a torch bulb (Figure 4.29b). (You may need one that is labelled 12 or 6 volts depending on the primary coils at your disposal.) Whether your transformer is square or buckle-shaped, your coil should be very loose so that it seems to have no contact at all with the transformer circuits. Then when it lights up in your hands the pupils really are surprised. Somehow the smart coils provided with the transformer do not seem to be really separate from the coil in the same way as a loose home-made coil does.

For class use, the Westminster Electromagnetic Kit can be used for making a transformer. This works far more reliably than the motor.

♦ *Further activities*

♦ Pupils could find out about the electrical systems of cars. They could look at a wiring chart in a repair manual and identify the alternator (generator), battery, rectifiers, fuses, etc.

♦ *Equipment notes*

♦ When looking at series and parallel circuits, it is often preferable to use a power pack, rather than batteries. This is because batteries tend to have high internal resistance. You may find that two bulbs in parallel draw less than twice the current of a single bulb, and so on. Alkaline-manganese cells have lower resistance than zinc-carbon cells, and rechargeable even lower.

♦ Large demonstration meters are available. These are excellent for demonstrating circuits to the whole class, since their scales can be read from a distance.

♦ You may have circuit boards available. These can be very convenient, and they allow pupils to set up circuits quickly, given a circuit diagram. But this very convenience can prevent pupils from thinking about the significance of what they are observing.

When asking pupils to set up a circuit for themselves, encourage them to lay out the components on the bench in the relative positions shown in the circuit diagram or drawing. Then they should wire them up, starting at one side of the power supply or battery and working round to the other.

♦ The Westminster Electromagnetic Kit contains all you need for class experiments involving magnetic fields, electromagnets, electric motors and transformers.

◆ *References*

ASE, 1995: *Signs, symbols and systematics: the ASE companion to 5–16 science*. ASE publications.

ASE, 1997: *World of Science*. John Murray/ASE.

Whitehouse, Mary (ed.), 1996: *Physics Activities for GCSE*. Nuffield Foundation, 28 Bedford Square, London WC1B 3EG.

◆ *Other resources*

- ◆ *Benjamin Franklin* by Nicola Kingsley, in the *Nature of Science* series, outlines how an early understanding of electricity was developed (ASE, 1989).

- ◆ Two useful CD-ROMs which allow pupils to construct, test and alter electric circuits on screen are: *Crocodile Clips* (see **www.crocodile-clips.com/education**), and *Edison* (REM Ltd, Great Western House, Langport, Somerset, TA10 9YU).

- ◆ *Wattville* is another CD-ROM; it looks at electricity consumption in the home and is suitable for lower secondary pupils (formerly available from Understanding Electricity).

- ◆ *The new way things work*, by David Macaulay, presents many electrical and electromagnetic devices in an entertaining way (Dorling Kindersley, 1998).

- ◆ For resources produced by the electricity supply industry, contact your local supply company, as there is no longer a unified education service.

Visits

It is possible to arrange class visits to power stations, and this can give pupils an opportunity to see turbines and generators at work. Many, particularly nuclear power stations, have visitor centres. Contact your local supply company.

Background reading

For a discussion of pupils' ideas of electricity and magnetism from a primary perspective, see: Bell, Derek, and George, Noel (eds), 1997: *Understanding Science Ideas: A Guide for Primary Teachers*, a *Nuffield Primary Science* publication. Collins Educational (HarperCollins).

For a simple explanation see: Sang, David: *Superconductivity*, an *Inside Science* supplement. *New Scientist*, 18 January, 1997.

A fascinating account of the way in which the nature of magnetism was unravelled is given in: Verschuur, Gerrit, 1993: *Hidden Attraction*. Oxford University Press.

5

Earth in space

Jonathan Osborne

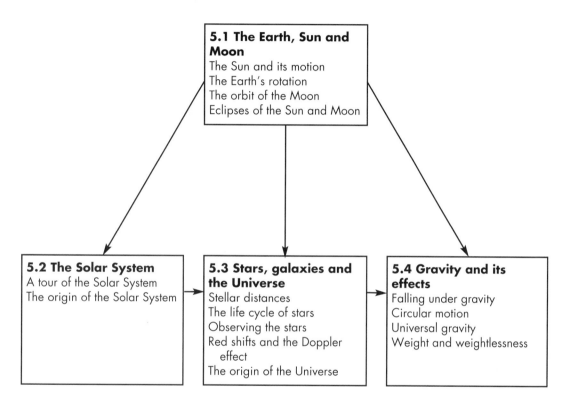

5.1 The Earth, Sun and Moon
The Sun and its motion
The Earth's rotation
The orbit of the Moon
Eclipses of the Sun and Moon

5.2 The Solar System
A tour of the Solar System
The origin of the Solar System

5.3 Stars, galaxies and the Universe
Stellar distances
The life cycle of stars
Observing the stars
Red shifts and the Doppler effect
The origin of the Universe

5.4 Gravity and its effects
Falling under gravity
Circular motion
Universal gravity
Weight and weightlessness

◆ *Choosing a route*

Typically, curricula are written so that they start with our local region of space (Earth, Moon and Solar System) and work outwards to larger structures (galaxies) and the entire Universe. Everyday observations are extended to evidence gathered by telescopes and space probes. Ideas about gravity are needed to explain many of these observations, and are probably best left until later in pupils' secondary studies.

♦ *Astronomy as a science*

Teachers often wonder why this topic is included in the National Curriculum as they feel that it is not part of the core sciences. This is a mistake. First, as well as things that are too small to see, such as cells and atoms, the natural world consists of objects that are unimaginably large or distant. Planets, stars and galaxies are such objects. Also, the story that we have to tell about the Earth, how old it is, how it came to be formed and how long it will last, is an important story which all children deserve to hear. For many of them astronomy is a point of fascination, as science fiction and space travel permeate much of their cultural experience.

Second, there is a limited amount of practical work associated with astronomy. This should be seen as an opportunity to use models, secondary sources and the Internet, which has a wide range of material on this topic. Pupils need to develop the skills required to read, analyse and summarise information from such sources and this topic provides such a challenge. The many excellent resources available can make the topic a visual treat.

Finally, our understanding of the Universe is an excellent example of the relationship between theory and evidence. Pupils commonly believe that scientists 'do experiments and find out things'. The view of science that emerges from studying our evolving ideas of the Universe is a very different one of individuals making observations and creating pictures of how the world might be, and providing ideas that can be tested. Thus work in this area can be very valuable for illustrating the relationship between theory and evidence and how our claims to know are justified in science.

Children's difficulties

There is a temptation to see much of this work as relatively simple and descriptive and not too conceptually demanding. After all, books seem to contain large bodies of factual information which children can reproduce rather mindlessly for assignments. However, this assumption is unwarranted. Table 5.1) summarises the conceptual transformation that has to be made between everyday or intuitive thinking and the scientific understanding. Everyday experience consistently reinforces the ideas in the second column. The contrast between intuitive and scientific thought shows that today's scientific understanding was the product of considerable struggle by some very creative minds. In this chapter, we will look at several useful ways of assessing whether pupils have really adapted their ideas to the scientific view.

Table 5.1 *A comparison of commonsense and scientific views about the Earth and space.*

Feature	Intuitive concept	Scientific concept
Size of solar objects	Earth is larger than the Sun and Moon which are larger than the stars	Stars are suns which are larger than the Earth which is larger than the Moon
Shape of the Earth	Earth is flat	Earth is spherical
Movement of the Earth	Earth is stationary	Earth rotates on its axis once every 24 hours and around the Sun once a year
Solar System	Rotates around the Earth (geocentric)	Rotates around the Sun (heliocentric)
Day and night	Sun moves, rising and setting	Earth spins, Sun stays still
Gravity	There exists an absolute 'down' which is the same everywhere	'Down' is towards the centre of the Earth and the direction varies across its surface.

A capital convention

Note that the names of specific astronomical objects are usually written with a capital letter (e.g. Sun, Moon, Mercury), but general terms are not (e.g. star, the moons of Jupiter).

5.1 The Earth, Sun and Moon

◆ *Previous knowledge and experience*

Children should arrive in secondary school knowing that the Earth, Moon and Sun are spherical, as well as the explanation for day and night and the period of rotation of the Moon.

◆ *A teaching sequence*

It is advisable to begin with some activities that check what ideas your pupils hold, before moving on to new ideas.

The Sun and its motion

Provide objects with the following shapes: a cylinder, a sphere, a hemisphere, a semicircular disc and a disc. Show them the shapes and then ask them to decide which shape most closely resembles the Earth, the Sun and the Moon. Research shows that some children of secondary age still think the Moon and the Sun are flat discs.

Now hand out a diagram similar to Figure 5.1 and tell them it is a view from a house looking South. Ask them to indicate the path of the Sun across the sky, and to mark where the Sun would appear at different times of day: as it rises, at midday in winter, at midday in summer, and as it sets. Research shows that many 11-year-old children do not have a clear conception of the path of the Sun across the sky during the day.

Figure 5.1
Looking towards the southern horizon. Note that this diagram is only relevant to observers in the northern hemisphere, north of the tropics.

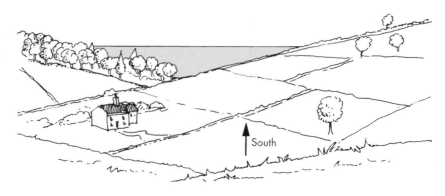

South

Give the pupils the following terms, in two columns:

Column A	Column B
Earth	day
Sun	night
Moon	year
	month

Ask them to write sentences that link items, including at least one from each column. Start by giving them an example which you should tell them is false, e.g. the Earth goes round the Sun in one night. Your pupils' responses will reveal areas of uncertainty in their knowledge; the activities that follow will help you resolve these.

Developing pupils' knowledge

Astronomy is very much an observational science. Pupils need to be asked to make observations and then possible explanations offered for discussion.

Ask the pupils to draw a picture of the view from the school, looking South. Ask them to mark where the Sun appears: when they come to school, in the middle of the day, and when they leave to go home. They can make observations before the next lesson, or draw their current ideas and compare their answers with their peers. Ensure that a well-understood consensus emerges.

Ask them to observe the stars. Get them to mark the position of a few prominent stars at the beginning of the evening and then two hours later. Do they appear to move? Is their movement the same as or different from the Sun's movement? (These observations may be difficult in some brightly lit locations.)

The Earth's rotation

Ask the pupils for explanations of what causes day and night. They should be able to suggest at least two: either the Sun moves around the Earth or (the scientific explanation) the Earth spins on its axis.

Model the explanations: ask the pupils to work in pairs. For the first of the explanations ask one pupil to imagine he/she is the Sun and another to imagine he/she is the Earth. The 'Sun' has to write instructions to the 'Earth' as to how it should move, and the 'Earth' has to write instructions to the 'Sun' as to how it should move. The pupils then pass their instructions to each other and act out the instructions. This activity should then be repeated for the scientific explanation.

How do we decide? Most children know that the spinning Earth is the correct explanation but very few can explain why. Two pieces of evidence are crucial. Firstly, a photograph or poster of the night sky taken with the shutter open for several hours shows that all the stars appear to rotate around one star in circles. (Such posters are available from the Armagh Planetarium – see *Other resources*, page 222.) Ask the pupils for possible explanations. There are two alternatives. Either all the stars are rotating around the one star, or the Earth on which the camera sits is spinning. How can we decide? The answer is that we cannot from this evidence alone. However, scientists do not prefer complicated answers. Which of these explanations is the simpler? On the basis that we prefer simple explanations we choose the idea that the Earth spins. This is an example of Occam's Razor that 'entities should not be multiplied beyond necessity' – put simply, one should choose the simplest explanation that fits the facts.

More substantive evidence for the Earth's rotation comes from the motion of a very long (over 10 m) and heavy pendulum. As the pendulum swings through the day, the line along which it swings gradually moves round as the Earth turns beneath it (Figure 5.2). Whilst this is almost impossible for schools to set up, it is worth mentioning that such a 'Foucault's pendulum' can be seen in a stairwell at the Science Museum, London. The effect was first demonstrated by Foucault in 1852 to Napoleon III at the Panthéon in Paris and was the first direct (non-deductive) evidence for a spinning Earth.

Figure 5.2
Foucault's pendulum.

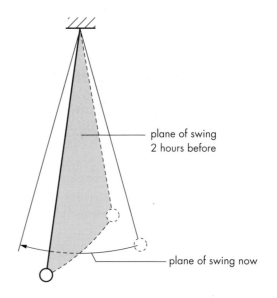

plane of swing
2 hours before

plane of swing now

The orbit of the Moon

Many children have not systematically observed the phases of the Moon over a period of a month, particularly urban children. Therefore this simple exercise is important. Children can each be given the responsibility for one night. Display a big chart on the wall which they can fill in. The book *Earth and Beyond* (ASE, 1997) contains a photocopiable chart.

The explanation of the changing appearance of the Moon is not easy because young pupils have difficulty imagining what the world would look like from somebody else's perspective. Start by sitting them in fours around an asymmetric object, such as a teapot or a cup, whose appearance depends on which side you view it from (Figure 5.3). Ask them first to draw what

Figure 5.3
Looking at an asymmetric object from different positions.

the teapot looks like to them, then to the person on the left, to the person opposite them, and to the person on their right. When they have finished they can compare their drawings with what the other pupils have drawn to see where they may have gone wrong.

Now repeat the exercise but this time use a large white ball illuminated from the side using an OHP. Arrange the class in four groups – directly in front of the ball, directly behind, to the right and to the left. Again get them to sketch what the ball looks like to them, and to the other viewers.

This exercise now has to be related to the view of the Moon that an individual will have at midnight when the Moon is the four positions shown in Figure 5.4. Ask the pupils to sketch how the Moon will look to the person on the Earth. Will the Moon be high in the sky, or low, near the horizon?

Figure 5.4
Positions of the Moon at new, half and full Moon.

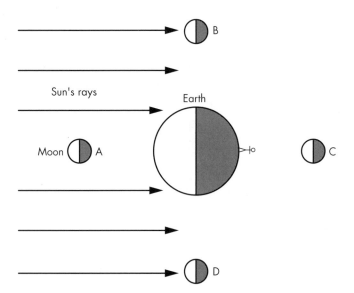

Why does the Moon keep the same side towards us? The reason is that the period of rotation of the Moon on its axis is exactly the same as the time it takes to go round the Earth once. This can be shown by using pupils as models. Ask one pupil to stand still as the Earth and ask another one to act as the Moon. Ask the 'Moon' to hold a large flat card upright on his/her head and move around the 'Earth', turning so that his/her face is always looking at the 'Earth'. Ask the children to watch the card and to work out how many times it turns around as the 'Moon' goes around the 'Earth'. (In one revolution around the Earth, the Moon turns once on its axis.)

Eclipses of the Sun and Moon

The more alert pupil may ask why the Moon is not eclipsed by the Earth when it is in the full Moon position shown in Figure 5.4. This is because the Moon orbits the Earth in a plane which lies at an angle to the ecliptic – the plane of the Earth's orbit around the Sun (Figure 5.5).

Figure 5.5
The plane of the Moon's orbit around Earth is tilted at 5° to the plane of the Earth's orbit around the Sun. Eclipses are therefore rarer than one might expect.

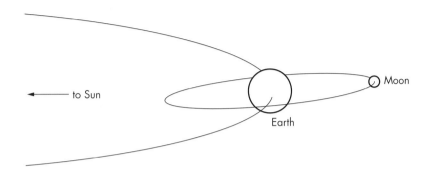

There are generally about two eclipses a year when the Moon blocks out the view of the Sun from the Earth. The path of totality in a solar eclipse is generally very narrow (up to 120 km wide), so few people are in a position to view it. The standard textbook diagram cannot be drawn to scale (the Sun is always too small compared to the Moon and the Earth, and far too near) and this is a point to be made to pupils. There is a wealth of information on eclipses on the internet (see *Other resources*, page 223) and it is even possible to view an eclipse 'live'. To see a total eclipse live is a dramatic experience as the following quotation shows.

> Then the moment of totality arrived. First there was the bright 'diamond ring' effect. Then there was a dramatic drop in light, the sky became nearly as dark as night. We could see stars in the daytime and most noticeably, several bright planets were clearly visible. People cheered and screamed and shouted, as if they were overcome by the spectacular nature of the event. There was a tremendous glow round the edge of the Sun that you would not imagine existed. This is the corona, it was the most beautiful blue colour and full of surprising movements of lines of light. The onlookers were shouting and crying, but at the same time very excited. It only lasts two minutes – I don't think I drew breath in that time – I just looked and wondered and felt amazing.

(GCSE student's account, *Total Eclipse of the Sun: Activity pack*, ASE, 1998)

An eclipse of the Moon, by comparison, can be viewed from half of the Earth's surface. These eclipses are less dramatic but they happen about once or twice a year so are worth pointing out to pupils.

> **!** *The Sun is naturally an intense light source. **NEVER** look at the Sun with the naked eye. Sunglasses do not provide sufficient protection and in the event of a solar eclipse, special viewing filters should be purchased. Check that they are not scratched or damaged any way. (Although there is a reflex to look away from bright lights this cannot be relied on, and in practice it may be ineffective as a result of nerve damage or drugs.)*

Viewing an image of the Sun with a telescope

A telescope or a pair of binoculars can be used to project an image of the Sun onto a screen using the arrangement shown in Figure 5.6. When this is done, small dark spots may be seen on the Sun. These are sunspots, cooler patches of the Sun's surface. The number of sunspots varies in an 11-year cycle.

> **!** *The previous warning is even more significant here. To look at the Sun through binoculars would produce instant blindness. Teachers must warn children that they must project the image, **NOT** look up the binoculars.*

Figure 5.6
Projecting the Sun's surface using binoculars.

5.2 The Solar System

◆ *Previous knowledge and experience*

Pupils will be familiar with a picture of the Solar System, with the nine planets orbiting the Sun. A more detailed discussion of the planets is now appropriate.

◆ *A teaching sequence*

Here are the principal ideas that we are trying to communicate to pupils.

- We live on a spherical object, a planet, which circles the Sun in what is called an orbit.
- There are eight other planets which do likewise.
- However, there are big differences between these planets. The inner planets close to the Sun – Mercury, Venus, Earth and Mars – are hard, rocky spheres and unusual in that their principal constituents are elements which make up only 2% of matter in the Universe.
- The next two planets, Jupiter and Saturn, are much further out, much larger and gaseous, consisting principally of hydrogen and helium.
- Then even further out still, are the outer planets – Uranus, Neptune and Pluto. These are denser than Jupiter and Saturn, but less dense than the Earth. It is thought that the interiors of Uranus and Neptune are composed of a deep ocean of water. Pluto and its companion Charon, on the other hand, both have a higher density, and are now thought to be rocky asteroids displaced from the belt between Mars and Jupiter.
- Many of these planets have moons which orbit the planet.
- Saturn has spectacular dust rings. Jupiter, Uranus and Neptune have much fainter rings.
- Our own planet is large enough not to have lost its atmosphere as there is sufficient gravity to prevent its escape; it is far enough from the Sun so that the surface temperature is such that water is a liquid and abundant life can survive. As far as we know, no other planet or moon in the Solar System has life on it.
- The Sun is a star which looks large because it is much nearer than the other stars. As stars go, it is slightly on the small side and about halfway through its lifetime of about 10 000 million years.

Put in that manner such information can seem almost mundane and trivial. Yet much of this information is contemporary science, only determined in the last 100 years and even, in some cases, in the last 20 years. Moreover, this knowledge is knowledge about our home – our place in the Universe – answering such questions as *where* we are, *what* we are and *who* we are, fundamental questions that have fascinated humanity since the dawn of time.

Learning from secondary sources

The problem posed by such a body of information is how to retain it. This is where it is important to employ a diverse repertoire of tasks which encourage children to assimilate this information. On pages 199–206 you will find a range of suitable activities – producing a concept map, research leading to production of a poster, using directed activities related to text (DARTs), constructing models, and so on.

A tour of the Solar System

For this exercise, a set of slides such as that entitled *The Solar System* (obtainable from the Armagh Planetarium, see *Other resources*, page 222) is most useful. Whilst there are some excellent videos, nothing quite captures the imagination as some of the beautiful slides that can be seen in this set. Moreover, a darkened room and a high quality, large photograph are much more breathtaking than even the best of videos. An additional extra is to make a tape of Holst's *Planets* suite and use this as an introduction to the show. As the teacher, you can ask your pupils to sit back and accompany you on a journey through the Solar System. More background information is available in the excellent book *Wanderers in Space* (Lang & Whitney, 1991); or visit the Armagh Planetarium web site **www.armagh-planetarium.co.uk** which has the scripts for their slide tours. Alternatively, pupils could take their own tour using a CD-ROM such as *Astronomy* (Anglia Multimedia, 1999).

It is not possible to give a detailed description of each planet here. You will want to discuss compositions, atmospheres, sizes, temperatures and so on. You will find some important points which you could incorporate in your commentary in the *Enhancement ideas* on page 206.

The origin of the Solar System

Where did it all come from? We can picture the Universe as a giant recycling system. Our Sun was formed from the debris left behind when earlier stars came to the end of their lives and exploded. The matter collapsed in under its own gravitational attraction, spinning as it went. The result was a central core which became the Sun, with dust and gas collecting in lanes around the Sun. The matter in the lanes then collapsed into spheres which gave rise to the planets. How do we know this when we were not there at the beginning? The answer is that we do not know for certain. However, astronomers can simulate the behaviour of matter in such conditions on a computer and such an outcome happens frequently, so we are fairly confident of the explanation. (This suggests also that there may be many more 'solar systems' around other stars.)

◆ *Activities*

Producing a concept map

In this activity, the pupils are given a set of terms relevant to the area that has just been discussed or about which they have been shown a video or slides. For instance, to produce a concept map of the Solar System, you could use the following set but for younger secondary pupils the number of terms should be kept to a maximum of ten. They are asked to work as a group and may need to be shown a piece of the concept map or a different concept map altogether in order to grasp the intent of the exercise (Figure 5.7, overleaf).

Terms that could be included in a concept map.

satellites	planets	stars
the Sun	craters	moons
Jupiter	meteors	hydrogen and helium
craters	atmosphere	shooting stars
comets	Venus	orbits
gravity	rings	Solar System

The following instructions should be given to pupils, working in groups of three or four.

• Read the list of terms and cross out any that mean nothing to you.
• Write each remaining one on a small square of paper.

- Place the squares on a larger sheet of paper; closely related terms should be placed close together.
- Discard any squares of paper that you are unable to fit in at this stage.
- Fix the squares on the sheet.
- Draw lines between related terms.
- On each line, write some words showing how the terms are related.
- Add additional terms if you find it helpful.

Figure 5.7
Part of a concept map for the Solar System.

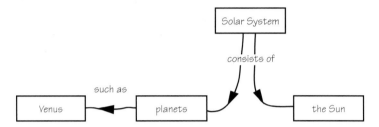

The activity normally takes about 30 minutes as first pupils have to discuss how to lay out the terms so that conceptually related ones are adjacent. When they have finished the task, they can then compare their map with others. This will show that there is no single right answer. The exercise provides a valuable opportunity for group recapitulation of the subject matter. Moreover, research shows that such activities do lead to a significant improvement in pupils' learning of science.

Producing a poster

Pupils can be divided into groups of three or four, each group charged with producing a poster about a different planet. They should be encouraged to find out answers to standard questions such as:

- How long is the planet's day?
- How long is the planet's year?
- What is the temperature at night?
- What is the temperature during the day?
- Does it have an atmosphere?

 Essentially, this should be offered as a research project and you will need a collection of resources including some CD-ROMs. When the pupils have produced their poster, it is very important to set aside at least five minutes for each group to present their poster so that the activity has a sense of purpose. Pupils make summary notes after each presentation. These can then be collated on a pre-printed table.

Tackling misconceptions

Another engaging activity uses a list of common misconceptions as a basis for discussion. Circulate a sheet containing ideas commonly held by pupils, such as the one shown in Table 5.2. Some of the statements are scientifically true and some are scientifically false. Pupils first have to work with the sheet on their own, going through the statements and deciding what they believe to be true (10 minutes maximum). When they have finished, they then have to work with two others, comparing what they decided and discussing why they decided what they did. Each group is charged with deciding what they think is an agreed answer (20 minutes). After this, it is important for you as the teacher to go through and explain what we believe to be the correct answer and how we know.

Table 5.2 *Commonly held ideas for discussion.*

Statement	Agree	Disagree	Don't know	Evidence
It is hotter in summer as we are nearer the Sun.				
The Earth is supported in space.				
The Sun generates its energy by burning coal.				
The Moon produces light in the same way as the Sun does.				
Gravity increases with height.				
From different countries we see different phases of the Moon on the same day.				
Astrology is able to predict the future.				
Planets cannot be seen with the naked eye.				

Using a spreadsheet

A set of data about the planets provides a very good example of the uses of spreadsheets for presenting and analysing data. Given data about the planets as a spreadsheet, pupils can be asked to plot:

• a bar chart of the distances of the planets from the Sun;
• a bar chart comparing the masses of the planets;
• a scattergraph of surface temperature against distance from the Sun.

(It is a good idea to ask them to decide whether a bar chart or scattergraph would be most appropriate.)

Using a formula for the volume of the planet they can calculate volumes in a new column. In the next column, they can then work out densities.

More useful spreadsheet exercises can be found in the book *Spreadsheets in Science* (Tebbutt & Flavell, 1995). The book has the added advantage that it comes with a complete set of spreadsheet templates on disk and the exercises are graded.

Making a wall model of the Solar System

At primary school, pupils may well have made a scale model of the Solar System on the school field. Here is another simple but effective exercise for conveying the relative sizes and distances of the planets. Pupils cut out a series of paper discs that are scale models of the planets. A good scale is 0.5 mm on the paper to represent one million metres on a planet, so that the diameter of the disc that represents Jupiter on this scale would be 60 cm and that for Mercury would be 2.4 mm. Alternatively, use a pre-printed sheet with a set of circles that represent the planets; see the book *Earth in Space* (Curriculum Council for Wales, 1991). A piece of string then needs to be hung across the wall and the planets pinned on at the appropriate position for their orbit. A good scale for this is 1 mm on the string to represent 10 million km in the Solar System. On this basis Pluto is 5.9 m from the Sun and Mercury 6 mm.

What this exercise shows most convincingly is the grouping of the planets, with the inner planets very close to the Sun, a large gap to Jupiter and Saturn, and then the outer planets that are much further away. Conventional diagrams in books tend to show the planets equidistant from each other which is a misrepresentation. More detail on this activity can be found in the book *Earth and Beyond* (ASE, 1997).

Quizzes

This is a topic which has a large amount of factual information associated with it. Learning facts can be made more entertaining by using quizzes and wordsearches. Examples of quizzes include solving anagrams using clues, e.g.:

| Most common chemical element in the Universe | DYHRGENO | _____ |
| Closest planet to the Sun | CRYRUEM | _____ |

or, producing a word description to be solved, e.g.

Temperature about 500°C, carbon dioxide atmosphere, pressure about hundred times that on Earth. Oh yes, and it rains dilute sulphuric acid! They called this planet after the goddess of love, but it seems a pretty hostile place to me.

Both of these examples are taken from the book *100 Science Puzzles* (Young & McCarty, 1992) which also contains crosswords and wordsearches.

◆ *Further activities*

Directed activities related to text (DARTs)

DARTs are based on the simple principle that reading scientific material is different from reading normal narrative text. Scientific reading has to be reflective, rather than receptive, and is an acquired habit. Therefore, it is no good expecting children to know naturally how to read scientific text. Instead, they have to be helped to sort and sift information and extract the salient facts. Space only permits the showing of two types of DART here but further information can be found in *Reading for Learning in the Sciences* (Davies & Greene, 1984).

In the first DART shown below, the really important activity is not so much the completion of the DART, rather the underlining and table construction which are the basis of extracting information from such pieces of extended text.

Shooting stars

Have you ever seen _____ stars flash across the sky on a summer night? No, they're not real stars that have got lost! They're bits of rock travelling incredibly fast through space. We call these objects *meteors*. When they _____ the Earth's atmosphere, the air _____ them down. Because they travel so _____, they get very hot and glow. They usually _____ _____ completely. All we see is the bright trail in the sky that lasts only a few moments. If the rock does reach the ground, it will make a _____ if it is big enough. We call these rocks _____. The planets are being hit by _____ all the time.

There are lots of craters made by meteorites on the Moon. You can see them easily through a small _____. The Moon doesn't have an atmosphere, so all the meteorites that cross its path hit the ground at full speed. Only the biggest meteorites land on _____. The others are _____ _____ in the layer of _____ surrounding the Earth before they get here. We're lucky that the Earth has an _____!

So, shooting stars (_____) are not very far away – just a few kilometres above our heads.

Instructions
- Fill in the gaps in the text above. Work in pairs and discuss each gap before filling it in.
- Underline in the text those items which describe the characteristics of *meteors*.
- Double underline in the text those statements which describe *meteorites*.
- Now make a table with two headings, *meteors* and *meteorites*. List the different properties of each.
- Finally write a sentence or two explaining the difference between a meteor and a meteorite.

In a scrambled DART such as that shown below, the sentence order is scrambled. The aim is to sort the text into a form where it makes sense. The activity requires the pupils to work in pairs and the text should be distributed on a printed sheet. The lines of text are then cut up and moved around. Again, cutting the text into strips is an essential part of the activity; trying to do the activity by merely numbering the order of these statements is much less effective. This DART, which concerns the importance of water on Earth, encourages discussion and thought about the science involved.

The density of water reaches a maximum at 4°C, so ice is less dense than water.
Any water on Venus would be in the form of steam, and water on Mars is now locked beneath the surface in the form of ice and frost.
As a liquid, it will dissolve almost anything to some extent.
As a result, ice floats on the surfaces of lakes and oceans, so they freeze from the top down.
Life as we know it, and the oceans of water that probably gave birth to life, exists only on Earth.
Ours is the only planet whose temperature matches the temperature of liquid water, between 0°C and 100°C.
The presence of liquid water makes the Earth unique among the known planets.
This provides an insulating layer that protects animals and plants from freezing.
Water is a marvellous substance.
We, ourselves, are largely water.
When we look at our nearest neighbours, we see that Venus is too hot, Mars is too cold.

Constructing a telescope

The range of traditional practical activities for this topic are limited. This fact should be used advantageously to emphasise that there are other ways in which people do science.

Making a telescope requires two converging (convex) lenses – a very thin lens (long focal length) for the objective and a fatter lens (short focal length) for the eyepiece, set up as described in Chapter 2 (see page 78). Few school lenses have focal lengths in excess of 50 cm and therefore it is difficult to make a telescope with high magnification. However, the exercise is still enjoyable as pupils like the opportunity to make a basic instrument.

Many pupils have difficulty seeing the image because the eye has to be positioned fairly carefully about 3 to 5 cm behind the lens. It is easier if they attempt to make a telescope out of two cardboard tubes (Figure 5.8). Both lenses are mounted with Sellotape or Plasticine, and the tubes can then be slid in or out until a magnified image is produced.

Figure 5.8
A simple telescope.

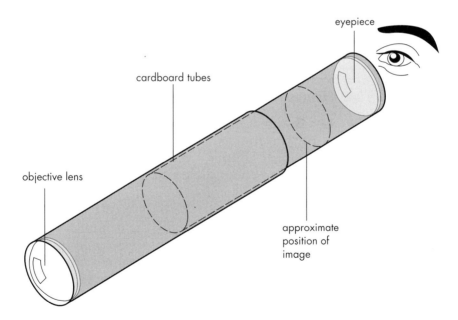

Schools often consider buying a telescope. The only telescope that can be recommended for school purchase is the *Astroscan* because of the high quality of the optics and its relative indestructibility (see the *Equipment notes*, page 221). Otherwise, a lot of good astronomy can be done with a reasonable pair of binoculars.

Modelling craters

This is a fun activity which illustrates how craters are formed. It is, however, rather messy so it requires lots of newspapers and a dustpan and broom. Craters are formed when a rock impacts with the solid surface of the planet. The best way to model this is to fill a tray with flour and smooth the surface by running a ruler across the top. Then drop a spoonful of flour onto the tray from a height in excess of 1.5 m. After several attempts, some of the results can bear a remarkable similarity to the craters on the Moon. (This does not work well if a large steel ball is used as the two substances are of very different densities.)

The activity can be taken further by measuring the depth of the crater and its width and comparing them with data for real craters from the Moon, where the ratio of width to height is typically 100:1.

♦ *Enhancement ideas*

Here are some points about constituent bodies of the Solar System; you might select from these to enhance a 'Tour of the Planets' slide show:

♦ The Sun consists of 73% hydrogen, 23% helium and 2% other elements. How do we know? Because the spectrum of the light emitted from the Sun is typical of these gases (rather as neon advertising signs give off their own characteristic colour).

♦ Ask how hot they think the Sun's surface is (6000 °C) and then how hot they think the central core is (14 million °C). Then ask what kind of process could release so much energy. Could it be the burning of coal or oil? If not, why not? The Sun is a wonderful example of a controlled nuclear fusion reaction in which hydrogen nuclei fuse to form helium atoms, releasing energy. If it gets too hot, the Sun expands and the reaction slows. If it gets too cold, the Sun contracts, warming the core and causing the reaction to speed up. (Inserting a photograph of an eclipse at this point would allow you to talk about eclipses, what causes them and what we can learn from them about stars like the Sun.)

♦ Some planets, such as Mercury, have cratered surfaces, the result of impacts by meteors or comets. Our atmosphere protects us from most impacts, but when one occurs it can create an enormous crater and throw up millions of tonnes of dust into the atmosphere, leading to dramatic cooling of the Earth's surface. Such an event is now considered to be the most plausible explanation for the extinction of the dinosaurs, as a massive crater has been found off the coast of Mexico, created about the time the dinosaurs disappeared. In 1995, there was a vivid demonstration of the collision of a comet with a planet when the comet Shoemaker-Levy collided with Jupiter; more recently, the possible collision of meteors with the Earth has been a theme of two films – *Deep Impact* and *Armageddon*.

♦ Venus is further from the Sun than Mercury, but hotter. Its atmosphere of carbon dioxide acts as a blanket, keeping it warm by the greenhouse effect. This shows what could happen to us if we continue to increase the levels of carbon dioxide in our own atmosphere.

♦ Photographs of the Earth taken from space, particularly the famous shot 'Earthrise' taken by the Apollo 11 mission, show it to be a small planet with its own atmosphere and predominantly covered in water. Some of that water is frozen. Again, ask how we know all this information. The series of slides entitled *The Earth from Space* (available from the Armagh Planetarium, see *Other resources*, page 222) illustrates the level of resolution that is possible with modern satellites and is a prompt for a discussion about the uses of such information.

♦ Show a slide of Mars, and put it out of focus. Can students see the 'canals'? This can lead to a discussion of how scientists often push their observational techniques to the limits and come up with misleading ideas.

   The probes that landed on Mars in the 1970s were designed to test for life; their results were inconclusive, but asking 'What tests would you use to test for living things?' is a good opportunity to remind pupils of the characteristics of living things and what might be relevant tests for the presence of life.

♦ The asteroids are the rocky remnants of a planet that never formed. The influence of Jupiter's gravity kept pulling them apart.

♦ Jupiter is like the Sun in that it is made predominantly of hydrogen and helium. In fact, if it had been twice as big, it might have become a star in its own right, producing energy by nuclear fusion. Then it is very unlikely that we would have been here, as there would have been two Suns orbiting each other.

♦ Saturn has the wonderful set of rings around it which make it the beauty of the Solar System. Until the first Voyager probe arrived to photograph it, we could only see three rings. The Voyager photographs show that there are hundreds of rings which consist of tiny pieces of rock and frozen gas in orbit around the planet.

♦ Uranus was the first planet to be discovered since ancient times, by William Herschel in 1781. Neptune was first observed in 1846; its discovery was a tribute to the power of maths and Newton's laws. Its position was predicted as a result of observations of variations in Uranus's orbit. These two planets give you a chance to talk about the importance of observation, instrumentation and calculation in astronomy.

♦ Pluto remains the last great mystery of the Solar System. It has only 1/440 of the mass of the Earth and has a companion Charon which was only discovered in 1978 and orbits Pluto once every 6 days keeping the same face towards Pluto. In Greek mythology, Charon was the name of the boatman who ferried new arrivals across the River Styx to Pluto's underworld. Penniless ghosts are said to have waited endlessly as Charon gave no free rides.

♦ Finally, it is worth mentioning comets, which are thought to be relics of the formation of the Solar System. They are in highly elliptical orbits around the Sun, hence they spend most of their time way out beyond Pluto and appear only rarely. The most well known is Halley's Comet which reappeared in 1986 and returns in another 76 years. (You could mention that some of your pupils will see it, even if you are unlikely to!)

♦ Meteors, in contrast, are particles of dust or small rocks that cross the Earth's path. When they enter the Earth's atmosphere, they are moving at very high speed and get red hot, giving off a brief but burning light which appears to be a rapidly moving star – hence the term 'shooting star'. Bigger meteors, which fail to burn up, hit the Earth's surface and cause craters, are known as meteorites.

5.3 Stars, galaxies and the Universe

No story about the Earth in space would be complete without some discussion of what we know about the life history of stars, how galaxies are formed, the origin of the Universe and why we believe it is expanding. Pupils are fascinated by the current scientific picture of a vast, expanding Universe, though it is difficult to convey more than a flavour of the evidence for this picture.

Again, the story that science has to tell to pupils is very dependent on secondary sources of information – a good collection of slides or one or two good videos are an absolute essential.

♦ *Previous knowledge and experience*

Pupils should be familiar with the idea that the Sun is a star, similar to many others seen in the night sky.

♦ *A teaching sequence*

The first point to make is that the distance to another star is large – very, very large indeed by human standards, so large that we talk about how long it would take a light ray, travelling at the speed of 300 000 km *per second* (emphasise *not* per hour) to get there. Using this as a basis for measurement, it would take eight minutes to get to the Sun and four years (!!) to get to the nearest star, Proxima Centauri. Since the fastest rockets only reach 28 000 km *per hour*, the idea of travelling to another star is impossible until somebody does manage to invent a warp drive!

For pupils who can make simple calculations of speed, distance and time, you could provide distances to some astronomical objects and ask them to calculate the time taken for light and for a rocket to travel these distances. Alternatively, give some distances in light-seconds or light-years, and ask pupils to suggest the objects they might refer to.

Stellar distances
The distances to nearer stars can be measured by *parallax* (Figure 5.9, overleaf). This is what causes a nearby star's position to appear to change against the background of the other stars between winter and summer. Since we know the distance between the two positions of the Earth, and we can measure

the angle by which the star has moved, it is simple trigonometry to work out the distance to the star, which is the height of the triangle. Note that the method of parallax only works for nearby stars.

Figure 5.9
The measurement of stellar distance by parallax.

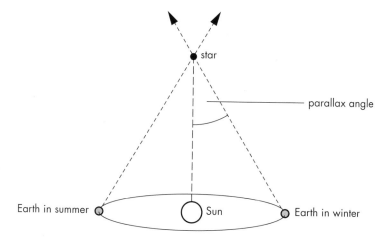

A concrete illustration of this effect can be undertaken outside, using one pupil as the star. Ask the other pupils to move from their position in summer to their position in winter and note how the 'star' has appeared to move against the background. Measure the angle between the 'star' and a distant tree, from each position. Now use scale drawing to find the distance of the 'star'. Check by measuring on the ground.

The life cycle of stars

The second point to make is that stars do not go on forever. Some of the stages that a star may go through in its life are given in the *Enhancement ideas* on page 215. When it comes to the end of its life, a star more than four times as heavy as our Sun blows up spectacularly in a supernova event. A smaller star gradually fades away to become a white dwarf and then a dark star.

The slide set *Stars and Galaxies II* (available from the Armagh Planetarium, see *Other resources*, page 222) contains some magnificent photographs, taken just two weeks apart, of a star that blew up in such a fashion. The Crab Nebula is the remnants of a star that was seen to do the same thing in 1054AD, an event which was recorded by Chinese astronomers. Pupils might look for this nebula, which is visible to the naked eye on a good dark night, lying below 'the belt' of 'the hunter' in the constellation of Orion, appearing as a small, fuzzy blob.

Observing the stars

The brightness of stars

Different stars have different colours, according to their surface temperatures. A yellow star (like the Sun) has a surface temperature of 6–7000°C. Others reach 20 000°C and appear blue. Surface temperature is very easy to measure from the spectrum – the hotter an object is, the bluer the light that emerges. You can show this with a 1.5 V bulb, controlled by a rheostat, connected to a 3 V battery. When it first comes on, the light is dim and red, but as it gets brighter, it becomes yellow and then almost white before it burns up.

How many stars?

Writing in 1929, this is the question that Sir James Jeans posed. His answer – that there are as many stars as there are 'grains of sand on all the beaches in the world' – is one way of conveying the enormity of the Universe and the number of stars that exist. It can be made into a numerical exercise by asking pupils to estimate the number of grains of sand in 1 cm³. (Divide the sand into, say, 20 equal-looking portions and count the grains in one.) Use this data to work out the number of grains on a beach, 1 km long, 50 m wide and 1 m deep.

Why do we believe that this is the right kind of figure for the number of stars there are? We know we live in the midst of a collection of stars. Looking at the night sky (in the countryside with no Moon present), you can see the collection of stars, edge on, which we call the Milky Way (Figure 5.10).

Figure 5.10
Our position in the galaxy.

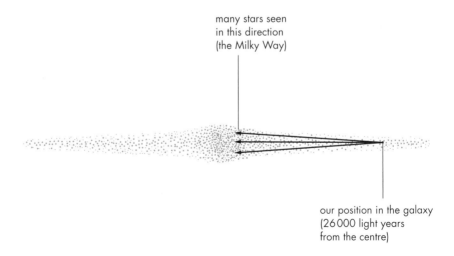

many stars seen in this direction (the Milky Way)

our position in the galaxy (26 000 light years from the centre)

Up until about 1925, we thought that only one collection of stars existed – the Milky Way which *was* the Universe. There were some funny, fuzzy whirlpool objects that looked as if they might be collections of gas and dust, but astronomers thought they were inside our galaxy. However, in 1925, Edwin Hubble turned the new giant, 200-inch Mount Palomar telescope on these and realised that they were themselves made of individual stars. These were other galaxies *outside* our own galaxy! Overnight, the Universe had suddenly become very much bigger; we now know there are thousands of millions of galaxies in the Universe.

Pupils enjoy writing very large numbers, and this can emphasise the scale of the Universe:

number of stars in a galaxy =   100 000 000 000
number of galaxies in the Universe =   100 000 000 000
so number of stars in the Universe =
                                  10 000 000 000 000 000 000 000

Red shifts and the Doppler effect

What Hubble did next was to look at the spectra of light coming from some of the stars in these galaxies. He found that, apart from some local galaxies, over 99% seemed to have their spectrum shifted towards the red.

This 'red shift' is an example of the *Doppler effect*. If a star is moving away from us, its spectrum is shifted towards the red; if moving towards us, it is shifted towards the blue. It is an apparent effect as it is a result of the source moving and not an actual change in the frequency of the light emitted by the star. Hubble concluded that nearly all the stars around us are moving away, which implies that the Universe is expanding.

Figure 5.11
Apparatus to demonstrate the Doppler effect.

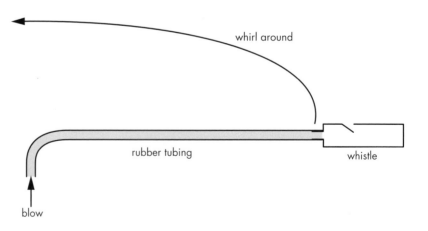

whirl around

rubber tubing

whistle

blow

The Doppler effect can be demonstrated with the apparatus shown in Figure 5.11. Attach a whistle firmly to about a metre of rubber tubing. Whirl the whistle fairly rapidly around your head while simultaneously blowing into the tube. The pupils should be seated in front. The effect is to produce a note whose pitch rises and falls. It rises as the whistle is moving towards the audience and falls as it moves away. An everyday observation of the same phenomenon is the drop in pitch of a police or fire siren as it approaches, passes and recedes into the distance. Light is similar – moving away from you the 'pitch' (frequency) of the light falls, shifting the spectrum to the red.

Pupils may not be familiar with the idea of spectra and spectral lines, but you can say that a yellow star looks redder if it is receding, and bluer if it is approaching.

An expanding universe
There is a natural tendency to assume that because everything is running away from us, we must be at the centre of the Universe. This would suggest our view is a privileged world view. You can show that if you live in an expanding Universe, the world appears to run away from you wherever you are. Take a balloon and, with a felt tipped pen, draw a series of galaxies on the surface of the balloon then partially inflate it (Figure 5.12). Alternatively, cut out paper galaxies and stick them on.

Position yourself in the middle of the class. Ask each pupil to focus on one galaxy and imagine this is where they are in the Universe. Then blow the balloon up more. Ask each one what has happened to the other galaxies, as viewed from their galaxy. There should be a universal answer that they have got further away, showing that expansion is observed wherever you are.

Figure 5.12
Using a balloon to show that an expanding universe looks the same from every point within it.

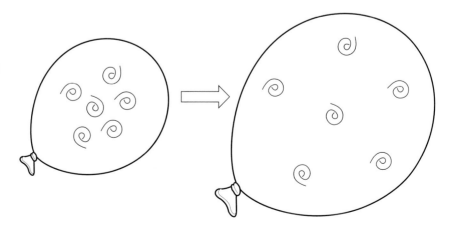

The age of the Universe

Hubble was able to work backwards to deduce when the galaxies must have all been packed closely together, shortly after the birth of the Universe. In this way, we can estimate the age of the Universe. Today's estimate is close to 15 billion years, though the figure has undergone many revisions in the light of new data in the past 20 years and is a point of some contention amongst astronomers.

Pupils could make a time-line to show the history of the Universe, starting with the Big Bang. The Solar System formed about 4.5 billion years ago, and life appeared perhaps 1 billion years later. They could also consider the future of the Universe. Current thinking is that it will expand for ever, becoming ever cooler and less dense.

The origin of the Universe

Hubble's finding suggests that the Universe started with a bang and has been expanding ever since; this is known as the 'Big Bang' Theory. Evidence for this explanation came in 1965 when two scientists, studying microwave radiation from astronomical sources, made the surprising discovery that wherever they pointed their aerial they picked up microwaves which corresponded to a very cool body of 2.7 kelvin (2.7 degrees above absolute zero). This radiation is the 'echo' of the 'explosion' at the beginning of the Universe. This theory is now believed in so strongly that one physicist was able to remark in 1985 that 'it is as certain that the Universe started with a Big Bang about 15 billion years ago as it is that the Earth goes around the Sun'.

Conflicting ideas

Note that, for some of your pupils, these ideas may conflict with ideas they receive from elsewhere, particularly religious ideas about the creation of the Universe. Whilst it is important to respect the right of individuals to hold such views, it should be made clear that the ideas you are putting forward as a science teacher are supported by a considerable body of scientific evidence gathered by telescopes and satellites and that amongst scientists there is no controversy apart from where the data is uncertain, such as for the Hubble constant and the mass of the Universe.

◆ *Enhancement ideas*

- ◆ A star starts life as a mass of gas and dust which collapses under its own gravitational attraction. As the matter falls inwards, it speeds up and becomes warmer. (Pupils can relate this to the work that they have done in developing a particle picture of solids, liquids and gases – faster particles mean hotter matter.) Eventually, the particles of dust and gas are moving fast enough to make the star just glow so that it is very red, but still very big. As it goes on collapsing, the interior becomes hotter, nuclear fusion begins and the surface of the star becomes yellow. The star then enters a long period of stability, in which the gravitational forces tending to make it collapse inwards are balanced by the outward pressure produced by the heat of the fusion reactions.

- ◆ Later in its life, a star may become a giant star or a dwarf star. Theoretical models, developed by astrophysicists, predict that when a star runs out of hydrogen to turn to helium, its interior will collapse under gravity and start to get even hotter. The temperature will then rise 20 or 30 million degrees. The outer layers of the star expand so that the star becomes much larger, and the outer surface of the star will get cooler and redder. The star will then be called a red giant. When this happens to our star, the Earth will be swallowed up.

- ◆ What happens next depends on the size of the star. Stars up to four times the size of our Sun will ultimately shrink under the pull of gravity to become small, white dwarf stars which gradually fade away. Larger stars go through a catastrophic event, becoming a supernova star, blasting their outer layers into space and leaving a core that shrinks under its own gravitational pull. The core goes on shrinking so that ultimately a matchbox of the matter that makes up the star would weigh as much as ten tonnes. Stars about six times larger than our Sun shrink still further so that the remaining matter becomes so dense that even light cannot escape from its immense gravitational pull. The resulting object is a black hole.

5.4 Gravity and its effects

◆ *Previous knowledge and experience*

Pupils will be familiar with the force of gravity, the name we give to the force that pulls things downwards towards the Earth's surface. In school science, they have to make the transformation from a world view in which 'down' is a direction between the two horizontal planes of the ground and the sky, to one in which 'down' is not absolutely fixed but is always directed towards the centre of the Earth (Figure 5.13).

Figure 5.13
The transformation between the commonsense view of 'down' and the scientific concept.

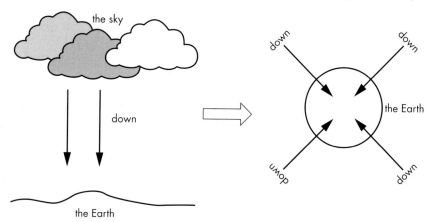

◆ *A teaching sequence*

Falling under gravity

Do heavier things fall faster?

Pupils are likely to hold the commonsense view that, the heavier an object, the faster it will fall. The video *Gravity* in the *Scientific Eye* series (from YITM) contains an excellent illustration of this phenomenon and shows young children dropping a range of objects off a high block. The real problem is that very light objects with a large surface area do fall more slowly because the air resistance slows them down more.

Demonstrate this with a stone and an unfolded piece of paper. Now scrunch up the piece of paper and ask them to write down which will fall faster. Most pupils will argue for the stone. You can safely bet them one Mars Bars each that there will be no difference! Releasing the two from a height of 2 m should prove this point.

Galileo's argument

Galileo came up with an argument to support the idea that objects of different masses fall at the same rate. His argument uses a dialogue between Simplicio (who is characterised as being a bit simple) and Salviatti (a know-all). Salviatti asks Simplicio to consider two stones, one heavier than the other, and asks, 'Which will fall faster?'. Simplicio states that it will be the heavier. Salviatti then poses the following conundrum:

> 'If we then take two bodies whose natural speeds are different, it is clear that on uniting the two, the more rapid will be partially retarded by the slower, and the slower will be somewhat hastened by the swifter . . .
>
>   'But if this is true, and if the large stone moves with a speed of, say, eight, while the smaller one moves with a speed of four, then when they are united, the system will move with a speed less than eight; but the two stones tied together make a stone larger than that which moved with a speed of eight.'
>
> (Quoted in Matthews, 1994, pp. 100–101)

In other words, a large object joined to a small object will fall more slowly, as the small object will slow down the large one.

The best way of using this is to give it to pupils and ask them to discuss it with the question 'Is Galileo wrong – can you mount a convincing argument to refute his argument?'.

The theory-based explanation is that heavier objects take more push to move. If one object is twice as big as another, the Earth will pull it with twice the pull, but there is twice as much mass to start moving so it will accelerate at the same rate as a lighter object. In effect, twice as much mass with twice as much pull results in the same effect.

The 'guinea and feather' experiment

The fact that it is the air resistance that makes a difference to the speed of fall of a feather can be shown by doing the famous 'guinea and feather' experiment. A glass tube containing a coin and a feather is connected to a vacuum pump (Figure 5.14, overleaf). The tube is first inverted while full of air. The coin reaches the bottom first. Now the air is pumped out. The tube is inverted and both coin and feather reach the other end at the same time.

Figure 5.14
Apparatus for demonstrating the 'guinea and feather' experiment.

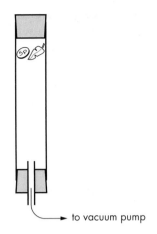

to vacuum pump

 This experiment should be done behind a safety screen and wearing safety glasses because there is a small chance of the tube imploding if there is a fault in the glass.

There are many videos and CD-ROMs showing Neil Armstrong doing this experiment on the surface of the Moon in 1969, without the need of a tube or vacuum pump. The advantage of using a CD-ROM is that the motion can be analysed graphically.

Circular motion

The point here is to take pupils through Newton's great insight – that motion in a circle is not a 'natural' form of motion and needs to be explained. The first point to make is that keeping something going in a circle requires a force and, most importantly, that this force is directed towards the centre. Both of the experiments to illustrate this are best done outside.

First, sketch out a large circle on the grass or playground, about 5 m in radius. Ask one pupil to stand on the circle and to walk in a straight line. (Make sure that, initially, the pupil is facing tangentially to the circle.) Ask another pupil to walk beside the first and to push or pull in whatever manner necessary to ensure that they follow the circle. This demonstration should make it apparent that they have to be pulled or pushed with a force which is directed inwards, *towards the centre* of the circle.

When you whirl an object on a string, you pull on the string to keep it moving in a circle. Removing this force does not result in the object flying radially outwards from the circle, but tangentially. A demonstration of this was suggested in Chapter 3 (see page 129).

Universal gravity

The point of these exercises is to lead up to the question that bothered Newton – if circular motion is not a natural form of motion, then what is the force that keeps the Moon going in a circle? Don't make the mistake of suggesting that 'Newton discovered gravity'. His imaginative leap was to extend the familiar idea of gravity as the force that makes the apple drop to the ground, by suggesting that it is the same force that keeps the Moon in its orbit round the Earth. The only reason that the Moon does not fall to the Earth is because the Moon is moving sideways.

It is useful to discuss Newton's 'thought experiment' here (Figure 3.18, page 131). A cannon ball fired from a high mountain falls to Earth. Fire it faster, and it travels further. At a sufficiently high speed, it keeps falling but its path follows the curve of the Earth. It is in orbit. The necessary speed is about 8 km/s.

Weight and weightlessness

Newton suggested that the Earth's gravity extended as far as the Moon, and indeed further. The Sun's gravity pulls on all the planets. An orbiting spacecraft is only slightly further from the centre of the Earth than we are, so gravity is only slightly weaker up there. So children may ask, why do people seem 'weightless' when they are in an orbiting spacecraft?

Weight is defined as the force of gravity on the mass of an object. Go up a mountain and you will weigh a little less, as does an astronaut in orbit around the Earth. However, the *sensation* of weight is the feeling that you get from the force of the ground supporting your weight. If the ground suddenly falls away beneath you, you begin to fall and feel weightless. The simplest way to experience this sensation is to drop off a diving board. You can also get a partial sensation in a lift when it slows down to stop after going up. In some lifts, as they start to accelerate downwards, you can sense a feeling of lightness as the force of the floor on your feet is reduced rapidly.

Astronauts in space appear weightless because they are in a 'box' which is also falling towards the Earth. Ask pupils to imagine standing in a lift when the cable breaks. Which will fall faster to the Earth – the lift or themselves? The answer is that they both fall at the same rate. Now ask them to imagine being in a box like the lift which is moving around the Earth and falling towards the Earth. If it is moving sideways at the right speed, it will keep its distance from the Earth and so remain in orbit.

In a lift or box that is falling freely, pushing with your feet on the floor will result in you 'floating' towards the ceiling. In reality, you are still falling but not quite as fast as the container. This is very difficult to illustrate but there are some good videos which demonstrate the effect. The video *Gravity* in the *Scientific Eye* series (from YITM) has a clip of a girl on a trampoline. As she starts to descend, she lets go of an apple and both she and the apple fall at the same rate so that the apple appears 'weightless'. It is reinforced by shots of trainee astronauts undergoing training in a jet aircraft which dives towards the ground like a falling object so that the individuals inside start to 'float around', apparently weightless.

Further activities

Putting objects into orbit: rockets

Putting anything into orbit means that work has to be done against the force of gravity to lift the object up. A lot of energy must be supplied to give an object sufficient velocity to get into orbit.

One good way of demonstrating rockets is to use a '*Rokit*' kit. These inexpensive kits (see page 134) convert a polycarbonate drinks bottle into a rocket which can reach a height of 100 feet. Pupils enjoy this demonstration. Even more dramatic are the plastic water rockets that can be found in toy shops. These reach greater heights and have the advantage that their release is controlled by the operator. Thus they are amenable to simple investigations which compare the height reached with the number of pumps of air supplied.

Enhancement ideas

♦ The reason that objects of different masses accelerate at the same rate under gravity is as follows. An object with greater mass is harder to accelerate, because of its greater mass. However, the pull of gravity on it is stronger, in proportion to its mass. The result is that double the mass experiences double the force and hence the same acceleration. So mass tells us about two things: how difficult it is to accelerate something (inertia) and how strongly the object is affected by gravity. No-one really understands why mass appears in both relationships.

♦ *Equipment notes*

The *Astroscan* Telescope, winner of an industrial design award, has a wide (3°) field of view which means that it is easy for children to locate the object. It does not need a tripod and its optics are sealed away from messy and dirty fingers. It is in a strong, plastic case and can be loaned to children. Its price is high, approximately £300 at the time of writing, but it is the only telescope that can be recommended. It is available from Ealing Scientific, Greycaine Road, Watford, Herts, WD2 4PW.

♦ *References*

 Anglia Multimedia, 1999: *Astronomy* (CD-ROM). Anglia Multimedia/ASE.

Association for Astronomy Education, 1997: *Earth and Beyond*. Association for Astronomy Education/ASE.

Curriculum Council for Wales, 1991: *Earth in Space*. Curriculum Council for Wales, Castle Buildings, Womanby Street, Cardiff, CF1 9SX.

Davies, F., and Greene, T., 1984: *Reading for Learning in the Sciences*. Oliver and Boyd (out of print but widely available in libraries).

Lang, K.R. and Whitney, C.A., 1991: *Wanderers in Space*. Cambridge University Press.

Matthews, M., 1994: *Science Teaching: The Role of History and Philosophy of Science*. Routledge.

 Tebbutt, M. and Flavell, H., 1995: *Spreadsheets in Science*. John Murray.

YITM: *Scientific Eye: Gravity* (videotape). ITPS Ltd, North Way, Andover, Hants., SP10 5BR.

Young, J. and McCarty, C., 1992: *100 Science Puzzles*. Collins Educational (HarperCollins).

◆ *Other resources*

- ◆ Look for details of phases of the Moon, eclipses etc. in diaries and the 'Night Sky' section of daily newspapers. You will find notes of other things to look out for in popular astronomy magazines such as *Astronomy Now* and *Sky and Telescope*.

- ◆ A useful resource is *Total eclipse of the Sun: Activity pack for secondary schools* (ASE, 1998).

- ◆ One of the best suppliers of educational resources for teaching astronomy is The Armagh Planetarium, Armagh, Northern Ireland. Materials can be ordered at very reasonable prices from:

  www.armagh-planetarium.co.uk/index.htm

- ◆ A very comprehensive resource list of posters, slides, videos, CD-ROMs, web sites and books, regularly updated, is available free from the Public Understanding of Science Team, Particle Physics and Astronomy Research Council, Polaris House, North Star Avenue, Swindon, Wiltshire, SN2 1ZZ.

- ◆ The following CD-ROMs are useful:
  Astronomy, Anglia Multimedia, Rouen House, Rouen Road, Norwich, NR1 1RB.
  Earth and Universe, BTL Publishing, Angel Way, Listerhills, Bradford, BD7 1BX.
  Encyclopaedia of Space and the Universe, Dorling Kindersley.
  Exploring our Solar System, Educational Media Film and Video, 235 Imperial Drive, Rayner's Lane, Harrow, Middlesex, HA2 7HE.
  Multimedia Motion, Cambridge Science Media, 354 Mill Road, Cambridge, CB1 3NN.

Visits

The Science Museum in London has a gallery devoted to space exploration; it also houses a Foucault's pendulum (see page 192).

Background reading
If you wish to extend your understanding of astronomy and how to teach it, you could look at the *Astrophysics Resource Pack* from the Teaching Resources Unit for Modern Physics, Science Education Group, University of York. This includes study notes on many aspects of astronomy and cosmology, together with teaching materials for older students.

Web sites
UK Eclipse web site
www.eclipse.org.uk
> A lot of information on eclipses is available at this site maintained by the Particle Physics and Astronomy Research Council (PPARC).

Bradford Robotic Telescope
www.telescope.org
> This site contains a wealth of resources for the teaching of astronomy in schools. It also permits data to be collected from the telescope for those who register with their scheme.

The Nine Planets
www.seds.org/billa/tnp/nineplanets.html
> A good and very extensive web site providing a wealth of detail about the Solar System.

The Royal Observatory at Greenwich
www.rog.nmm.ac.uk/astroweb/index.htm
> This is the site run by the Royal Observatory and contains many good educational tours of the Solar System.

Hubble Space Telescope
oposite.stsci.edu/pubinfo
> This site provides many of the latest facinating pictures from the Hubble Space Telescope.

NASA
www.nasa.gov
> Access to the wide range of resources provided by NASA.

6 *Radioactivity*

David Sang

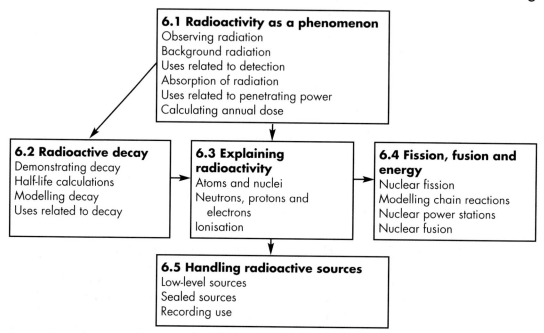

6.1 Radioactivity as a phenomenon
Observing radiation
Background radiation
Uses related to detection
Absorption of radiation
Uses related to penetrating power
Calculating annual dose

6.2 Radioactive decay
Demonstrating decay
Half-life calculations
Modelling decay
Uses related to decay

6.3 Explaining radioactivity
Atoms and nuclei
Neutrons, protons and electrons
Ionisation

6.4 Fission, fusion and energy
Nuclear fission
Modelling chain reactions
Nuclear power stations
Nuclear fusion

6.5 Handling radioactive sources
Low-level sources
Sealed sources
Recording use

◆ *Choosing a route*

This chapter starts by considering the phenomena of radioactivity: radiation and radioactive decay. It then looks at explanations at the atomic and nuclear level. Finally fission and fusion are considered.

Pupils may have already encountered ideas about atomic structure – perhaps in their study of chemistry – and you may then wish to reinforce these ideas by referring to them during your consideration of radioactive phenomena.

The usual laboratory regulations and considerations of good practice apply to work with radioactive materials. However, there are also additional regulations specific to the handling of radioactive materials in schools, and you need to be aware of these. They are dealt with in the last section of this chapter.

When working with pupils up to the age of 16, you, the teacher, are the only one allowed to handle most radioactive sources, and so many experiments have to be demonstrated. However, pupils can observe, record and process results, and there are also many simulations and other activities to keep pupils active.

6.1 Radioactivity as a phenomenon

Radioactivity has a mysterious quality to it. It is all around us, and yet our bodies have no organs for sensing it. (This may suggest that it hasn't been very important for us in our evolutionary history.) We have only known about radioactivity for about a century, since its discovery by Henri Becquerel in 1896.

Pupils make jokes about radioactivity – it makes you grow two heads, you will glow in the dark, and so on. It is worth acknowledging this from the start. A theme for this first section of study might be: Just how dangerous is radioactivity? How can we deal with it safely? Then this can be extended to: How can we make use of radioactive materials? If you can emphasise how you are treating radioactive materials with appropriate caution, you can help pupils to reach a realistic perception of radioactivity, its hazards and its benefits.

♦ *Previous knowledge and experience*

Most teaching about radioactivity occurs in the later years of secondary education. Pupils may previously have come across the use of radioactive materials (uranium and plutonium) as nuclear fuels in power stations. This may have introduced them to the ideas of the hazards of radioactive materials, and their decay.

♦ *A teaching sequence*

This sequence starts with three different ways of observing radiation from radioactive substances. Clearly, your choice depends on the equipment available to you. However, it is preferable not to start with a Geiger counter, since the other suggested techniques can give a more dramatic and concrete introduction to the topic.

Observing radiation

The cloud chamber

The simplest version of the cloud chamber is excellent for this purpose. The felt in the chamber is soaked in alcohol, and vapour then fills the chamber. Dry ice (solid carbon dioxide) in the base cools the vapour so that it becomes super-cooled – it wants to condense. Radiation from the source (a blob of radioactive paint) ionises the air and the alcohol vapour condenses around these ions. Tracks of tiny droplets radiate out from the source. It can help to rub the lid of the chamber with a dry cloth, as the static electricity generated helps to clear away any old tracks. (You can compare these tracks to the vapour trails of high altitude jet aircraft, where particles in the exhaust gases nucleate the formation of ice crystals.) It is extraordinary to realise that, with each track, we are witnessing an event at the level of a single atom.

Points to make:

- We can't see individual particles coming from the source, but we can show up their paths using this clever device.
- The tracks are more or less straight lines, radiating out from the source.
- The tracks appear sporadically – radioactivity is a random process.
- The particles travel only a few centimetres in air at atmospheric pressure.
- Occasionally, a track will be seen which cuts across the chamber; such tracks arise from cosmic rays, radiation from space.

A school cloud chamber is an unsophisticated version of the various cloud and bubble chambers and other detectors which are used today in high energy particle physics research, and which produce the striking images often used to illustrate articles about fundamental particles.

Take care with dry ice – use thick gloves or tongs to handle it so it does not come into contact with bare skin. It is a good idea to extract enough CO_2 from the cylinder to last the whole lesson, break it up into small chunks, and store these in a wide-necked vacuum flask, rather than trying to extract small quantities individually.

The spark detector

This is another dramatic demonstration. A spark detector (Figure 6.1) consists of a fine wire connected to about +5 kV and mounted a few millimetres below a wire mesh at 0 V. You will need to have access to a 5 kV EHT (extra high tension) supply for this demonstration. Start (without a radioactive source) by turning up the voltage until sparks jump between mesh and wire, like miniature lightning flashes. Ensure that pupils can see the sparks. These result from the presence of ions in the air which allow it to conduct. Turn down the voltage until the sparks just stop. Now, bring a radioactive source (of alpha particles) downwards towards the grid. When it is a few centimetres from the grid (depending on the source), the detector will start to spark.

Figure 6.1
A spark detector.

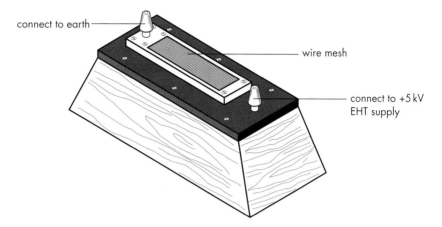

connect to earth

wire mesh

connect to +5 kV
EHT supply

Points to make:

- Again, we are showing up the invisible radiation from a radioactive source.
- Withdrawing the source shows that the radiation travels about 5 cm through air.
- Turning the source sideways stops the sparks, because the radiation is no longer directed downwards to the detector.
- A sheet of paper will block the radiation – this is the first point at which you can demonstrate the absorption of radiation.

You may also wish to show that there are radioactive sources which do not affect the spark detector. This is because it (and the cloud chamber) are best for showing up alpha radiation; beta radiation can be seen in a cloud chamber, but in schools a Geiger counter is needed to detect gamma radiation. However, at this point, it is unnecessary to introduce the terms alpha, beta and gamma.

The Geiger counter

You can now introduce the Geiger counter. Point out its components: the tube, where radiation is detected, and the box of electronics which does the counting. It may give a digital display of counts, or an indication of the rate (number of counts per second). It is most helpful to allow pupils to hear the detector at work; if it has no built-in loudspeaker, then connect it to an external speaker. A volume control is very desirable. You may wish to describe the tube as being like a spark detector: there is a metal rod inside (which corresponds to the wire), and the metal case is the like the grid, rolled into a cylindrical shape. Some radiation (gamma) can pass through the case. Some can only enter through the end window. Warning: the window is very fragile – do not touch.

Try some or all of the following:

- Hold the tube near the box of sources. You will hear a strong count.
- Hold it near some samples of radioactive rocks, and observe that naturally occurring materials are radioactive, some more than others.
- Listen to the background radiation in the laboratory (see opposite).
- Listen to the radiation from a beta source.
- Try blocking the radiation from these different sources with paper – can you detect a decrease in the count rate? (More about absorption shortly.)

Some points to highlight

It is important to distinguish between a radioactive source or substance and the radiation which it produces. This is similar to the distinction between a lamp and the light which it emits, or a radio transmitter and the radio waves it emits. A simple diagram (Figure 6.2) illustrates this; pupils should be able to give other examples of source and radiation (e.g. heat from a fire). Similarly, the term *radioactivity* should be reserved for the phenomenon itself, and not used as a shorthand for the radiation.

Figure 6.2
Comparing a radioactive source with a light bulb.

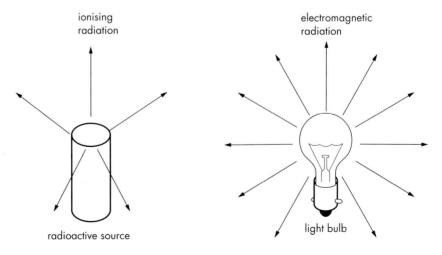

ionising radiation

electromagnetic radiation

radioactive source

light bulb

All the detectors described above rely on the *ionising* properties of radiation. How much you say about this depends on what your pupils already know about atoms, atomic structure (nucleus and electrons) and ionisation. You may decide to leave this point until later, or it may be enough to describe the radiation as 'crashing into the atoms or molecules of the air and smashing them up'. They become ions, and these are what we detect.

Although some laboratory sources may seem very powerful, this is something of an illusion. A Geiger counter is an extremely sensitive instrument. Each click represents the emission of radiation by a single atom. (Would eating a single atom of arsenic poison you?) You can take this point further shortly when discussing the question of the hazards of radiation.

Background radiation

Now that we have seen how to detect radiation from radioactive substances, we can start to think about its nature. Listening to the background count rate on a Geiger counter can be very instructive. Firstly, the rate is low – perhaps one count every second or two, depending on your geographical location. In future measurements, it may be necessary to take account of background radiation. Secondly, it is sporadic and unpredictable. Can you detect any rhythm? No. One consequence of this is that you need to measure over a significant period of time (perhaps five minutes) to determine the *average* background rate. Try recording the number of counts in successive intervals of 10 s, perhaps using a datalogger. How much do they vary? (More than you might expect!)

You might now comment on the question of safety. We are bathed in background radiation all our lives. Any exposure to radiation is, to some degree, hazardous. However, we appear to be able to tolerate background levels. So perhaps a click every second or two on the Geiger counter represents a reasonably safe level. To set an acceptable level of exposure, we might agree that exposure to radiation at work or in the environment should not increase our dose much above this level. (But bear in mind that background levels vary by at least a factor of three in the UK.)

You will want to consider the exposure of pupils and yourself in the course of this work. Using a Geiger counter, determine how far you have to be from a source for its radiation level to be low, close to background levels. You will probably find that, for a gamma source, you are safe beyond a distance of about a metre. Hence pupils should be kept at least a metre from the source, and you should work at this distance from the source, except when you have to manipulate it.

Uses related to detection

Radioactive substances have applications because their radiation can be detected at very low levels.

- Environmental tracers are used for determining the movement of substances such as water and pollutants.
- Biochemical tracers are radioactively labelled substances which are put into the body. Their movement is then followed. Some images of brain activity are obtained like this.

Absorption of radiation

It is desirable to mount both the radioactive source and the Geiger tube in holders as shown in Figure 6.3, so that they remain safely in place, and you can then work at a safe distance from the source.

Figure 6.3
Investigating absorption. (Adapted from: Basic Physics 1 and 2 by UCLES, 1995. Cambridge University Press.)

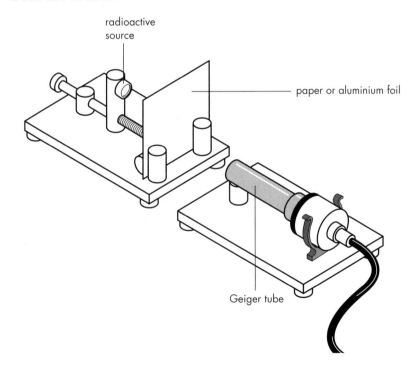

radioactive source

paper or aluminium foil

Geiger tube

At this point, it is best to discuss the three types of radiation produced by laboratory sources. The difference between them can be shown by their different penetrating powers.

1. With an *alpha source* in position, move the Geiger tube towards it. (You will need a tube with a thin end window to allow alpha radiation through; alternatively use a solid state detector.) Show that the radiation is only detected within a few centimetres of the source.

    With a gap of 1 cm between source and tube, place a sheet of paper between them. The reading will drop to background level. (If it doesn't, try two sheets. If you still detect an above-background level, your source is probably emitting beta and/or gamma radiation as well as alpha.)

    Try a much thinner sheet of paper, such as very thin tissue. Some radiation may pass through.

    Show that aluminium foil and lead also absorb all of the alpha radiation.

2. Replace the alpha source with a *beta source*. (If two beta sources are available, choose the weaker because of the high sensitivity of the Geiger counter to beta radiation.) Show that paper has little absorbing effect.

    Show that one thickness of aluminium foil has more effect. Try increasing thicknesses of foil, until the radiation level approaches background level. (You may need to use aluminium sheets.)

3. With a *gamma source*, show that aluminium foil has little effect; thicker aluminium has more effect, and a few centimetres of lead are needed to reduce the radiation to a level close to background level.

Now we can characterise the three types of radiation by their penetrating power:

*   alpha is the least penetrating;
*   gamma is the most penetrating;
*   beta is 'in between'.

You may wish to relate this to the speeds of the different types of radiation. Alpha travels most slowly, so that it has the most time to interact with the atoms it passes. Gamma travels at the speed of light, so that it can whizz through materials with very little interaction.

The absorbing power of materials can be related to their density; in simple terms, a dense material (such as lead) is more absorbing than a lightweight material (such as paper).

Note that pupils may have already come across gamma radiation if they have studied the electromagnetic spectrum. You can describe gamma rays as being 'the same as X-rays, but more energetic'. (The distinction is really one of origin: gamma rays are produced by radioactive sources; X-rays are produced when electrons decelerate.)

Storing sources

You could discuss the ways in which radioactive sources are stored in terms of radiation penetration. A lead-lined box is important for absorbing gamma radiation, but you will probably have found that a Geiger counter registers a significant level of radiation outside the box. Hence the box should be stored in a locked cabinet somewhere where people are not regularly in close proximity.

Uses related to penetrating power

Radioactive substances have applications because of the penetrating power of their radiation.

- Thickness detectors are used in paper and polymer factories. Beta radiation is absorbed by the manufactured material; if the material is too thin, more radiation gets through to the detector.
- Domestic smoke detectors have an alpha source in them. Smoke absorbs the radiation and this triggers the alarm.
- Gamma radiation is used to check for cracks in solid objects. It may be detected photographically.
- Medical equipment is irradiated with gamma radiation to sterilise it, as the radiation kills potentially harmful micro-organisms.

Pupils should be able to explain the choice of radiation in each case. Sterilisation by gamma radiation is more generally used than pupils may be aware. As well as being used for medical items such as syringes, which are sealed in their packaging before sterilising, the technique is also used for everyday items such as toothbrushes and tampons.

Calculating annual dose

Data for calculating pupils' approximate annual exposure to radiation is available in various forms: one convenient place to find an exercise is *World of Science* (ASE, 1997). Pupils can estimate their exposure to radiation from various natural and artificial sources. They should find that artificial sources account for a small fraction of their exposure, and that this proportion is small compared to the variation of background radiation in different parts of the country. From this we can conclude that exposure to radiation from artificial sources (mostly medical X-rays) is relatively insignificant, though we should still try to minimise this.

It is very hard to give a good estimate of the effect of background radiation on human health. It seems likely that the death toll from radiation-induced cancer is similar to the death toll on the roads. It is much less significant than the effects of smoking. Pupils could discuss the extent to which these risks are comparable. (Which are voluntary and which involuntary?)

By calculating their exposure ('dose'), pupils will learn that the risk from radiation is cumulative. It increases with the level of radiation and the duration of the exposure. Exposure from all sources must be added together. They will also find that radon (from the ground and building materials) gives the greatest contribution to their exposure. In fact, radon is thought to be the greatest health hazard in the natural environment.

Some points to highlight

You can describe the health hazards of exposure to radiation in terms of the radiation 'smashing into' the delicate cells of our body. Later, when ionisation is explained explicitly, you can give a better picture.

It is worth re-emphasising the distinction between the radioactive substance and the radiation that it produces. Radiation penetrating our bodies can harm us; a radioactive source inside us (such as uranium ore dust) can be much more damaging. The layer of dead skin on our bodies is sufficient to protect us from harmful external alpha radiation.

◆ *Further activities*

It is unlikely that pupils will carry out extended investigations in this topic, because of the restrictions on the handling of radioactive sources. Suggestions for further activities include:

◆ Discuss where pupils have seen radioactivity warning symbols.

◆ Use *Tastrak* detector strips to detect radon. After development, these show the tracks of radiation from atmospheric radon. Full details are given in *SATIS* Unit **1105b** *Radon – an investigation* (ASE, 1991).

◆ Re-enact the original experiment of Becquerel. Detailed suggestions are given in *Henri Becquerel and the Discovery of Radioactivity* (Sang, 1997).

◆ Practise writing the Greek letters α, β and γ. Pupils need help to become familiar with these.

◆ Carry out a survey of public knowledge of and attitudes to food irradiation.

 ◆ Use a Geiger counter and datalogger to record background radiation levels over a period of days. You may find that radon levels build up at night (because there is little air circulation), giving enhanced readings.

◆ *Enhancement ideas*

◆ Radioactivity was discovered in 1896, shortly after X-rays. Becquerel had joined in the search for other forms of invisible radiation.

◆ The name 'radioactivity' was coined by Marie Curie, who discovered polonium and radium in 1898.

◆ In the early years of the 20th century, all kinds of radioactive substances were promoted as health products, including radioactive toothpaste, cosmetics and drinking water!

◆ In some areas, medieval miners died from lung cancer caused by radon gas which accumulated in the mines. One woman had a succession of seven husbands, because the death rate was so high.

◆ Even today, some European spas include exposure to radon-rich air in old mine workings as part of their treatment.

6.2 Radioactive decay

In the previous section, the random nature of radioactive decay was pointed out. Now we can go on to consider the typical pattern of decay which emerges from this underlying randomness.

◆ *A teaching sequence*

This sequence starts with a demonstration of radioactive decay, and then looks at ways of simulating decay before considering its implications. Note that there is no need at this stage to discuss what is going on in terms of atoms and nuclei; this is dealt with in Section 6.3.

Demonstrating decay

Using a protactinium generator

Demonstrating radioactive decay is bound to be tricky. You need a material which will decay during the course of a lesson, so it should have a half-life of a few minutes. If you order such a substance from the suppliers, there won't be much left by the time it reaches you! What is needed is a continuous supply of such a material, together with a way of extracting a sample for use when you need it.

The protactinium generator is the best way of achieving this. We will first consider how to demonstrate the decay of protactinium, and then look briefly at how the generator works.

Clamp a Geiger tube pointing vertically downwards (Figure 6.4, overleaf) so that the protactinium generator, standing in a tray, will slide neatly beneath it, the closer the better.

Ideally, the Geiger tube should be connected to a counter which can count automatically for ten seconds, display the result for a brief interval, and then start counting again. A suitable counter is the *Digicounter* from Philip Harris.

Wear disposable plastic gloves.

Shake the sealed bottle vigorously for half a minute. Place the bottle on its tray and slide it under the Geiger tube. Start the counter.

Now, simply observe the rate at which the numbers change on the counter (or listen, via a loudspeaker). After a minute or two, the rate will decline noticeably. The protactinium is decaying.

Figure 6.4
*Measuring the
half-life of
protactinium-234.
(Adapted from:
Basic Physics 1
and 2 by UCLES,
1995. Cambridge
University Press.)*

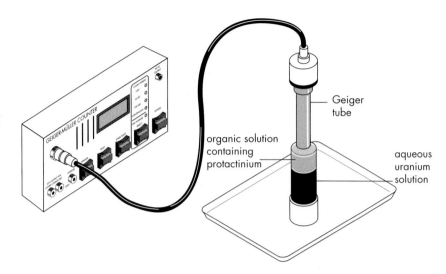

Geiger
tube

organic solution
containing
protactinium

aqueous
uranium
solution

To collect data suitable for analysis, shake the bottle again and
return it to its position under the Geiger tube. Record the
reading that is displayed for each 10 s interval; continue for
about five minutes. (The class can record the data for
themselves.) Convert each reading to a number of counts per
second.

You need to take care when plotting a graph of the results. If
you use a *Digicounter*, each data point represents the number
of counts in a 10 s interval, but the data points are separated by
13 s. The result of graphing the data is a somewhat erratic
curve, but radioactive decay is clearly evident. Find a point near
the top of the downward slope; find the point that is at half
this height. The time difference between these points is the
half-life of protactinium – about 72 s. Repeat for several
different starting values.

An alternative approach would be to use a ratemeter and
datalogger. This can present a real-time graph as the
experiment proceeds; a printout can be photocopied for
analysis by each pupil (Figure 6.5). It helps to start with a
dummy run, during which you remove the source. The screen
will show an immediate response, convincing the class that they
are truly observing something that is going on in the plastic
bottle.

How the protactinium generator works
There are two liquids in the generator, an organic solution
floating on a denser aqueous solution. The generator contains
a uranium salt, dissolved in the aqueous solution.

Figure 6.5
*A radioactive
decay graph.*

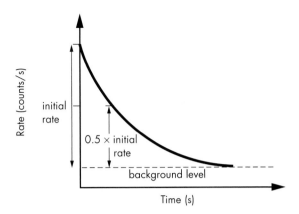

Protactinium-234 is produced when uranium decays, so
protactinium is being continuously produced in this lower
layer. However, it is also continuously decaying, so the amount
available reaches a fixed level where the rate of production
equals the rate of decay.

To separate the protactinium from the uranium, you shake
the bottle. The protactinium dissolves preferentially in the
organic layer at the top of the bottle, and it is the decay of this
sample of protactinium that you observe in the experiment.
Meanwhile, protactinium is regenerating in the lower layer,
ready for the next demonstration. Because you can repeatedly
extract protactinium from it, the bottle is sometimes known as
a 'protactinium cow'.

Protactinium emits beta radiation as it decays. The bottle is
made from a plastic which allows most of the beta radiation
from the top layer to pass through and reach the detector. Beta
radiation from the fresh protactinium in the lower layer is not
detected with the set-up shown in Figure 6.4.

 *The level of radiation from this source is low, provided it remains sealed.
However, do not spend a lot of time close to it. Always wear disposable
gloves in case of the unlikely event that the sealed bottle leaks while you
are shaking it. Stand the bottle in a tray, for the same reason. Return it (in
its tray or other container) to the locked storage cabinet immediately after
use.*

Half-life calculations

Once pupils have seen a decay curve, you should emphasise its
characteristic features. Pupils often assume that, since half the
material decays during one half-life, then all of it will decay in
twice this time. Indeed, it is often talked of in this way in the
media, for example 'the half-life of this radioactive waste is a

million years, so it will have to be stored for two million years'. From the shape of the decay curve, this is clearly untrue. After two half-lives, one quarter will still remain, and so on.

Pupils should be able to answer simple questions, such as:

'If the half-life of a radioactive substance is 3 days, what fraction will remain after 9 days?'

This is most easily approached stepwise, calculating the fraction remaining after 3 days, then 6 days, then 9 days. (Avoid setting questions where the half-life is a multiple of two; too many twos and halves can get confusing.)

Figure 6.6
Diagrams like this can help with half-life calculations.

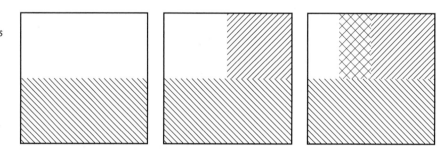

Activity and becquerels

The number of undecayed atoms in a sample decreases in this characteristic way, and so does *activity*. Activity is a property of a radioactive source; it is the number of decays which occur per second and is measured in becquerels (Bq). 1 Bq is one decay per second.

It is difficult to measure directly the activity of a sample (because you would have to detect every bit of radiation emitted). However, pupils should appreciate that, if a sample has an initial activity of 100 Bq, this will decrease to 50 Bq after one half-life, and so on.

Modelling decay

Pupils can enjoy taking part in whole-class modelling of radioactive decay, and get more of a feel for its nature. These activities underline the essentially random, spontaneous and unpredictable nature of radioactive decay. We cannot predict when an individual atom will decay, but we can describe the general pattern that emerges.

♦ **Model 1.** Each pupil requires a coin; hopefully they can provide their own. Everyone stands up, to represent undecayed atoms. Record how many there are. The coins are spun; anyone with a tail sits down as they are deemed to have decayed. Count how many remain undecayed. Repeat the process through several cycles. Plot a graph with the points equally spaced along the time axis. This will generate a roughly exponential decay curve, but it won't be perfect. The randomness of coin-tossing ensures that. You can discuss why, if you started with a class of 32, the numbers won't simply go 16, 8, 4, 2, 1. Repeat the experiment and see that you get different numbers but the same sort of pattern.

♦ **Model 2.** Provide each pupil with a single die. Those throwing a six have 'decayed'.

♦ **Model 3.** Use a large number of wooden cubes, each with a black dot on one face. Roll them at random. Extract all those with their dot uppermost – these have decayed. Roll the remainder, and so on. (If possible, have a second set, each of which has black dots on two faces. Ask the class to predict how their decay will differ from the first set.) If you cluster together the cubes that decayed the first time, the second time, and so on, you will see that they also display the same pattern.

With these models, it is desirable to discuss their limitations. Atoms do not disappear, as some pupils will imagine, when they decay. They are simply no longer part of the undecayed population and cannot emit radiation again (at least, not in these simple models).

Pupils may have difficulty in appreciating that both the population of undecayed atoms and the rate at which they decay follow the same pattern. You can discuss this as hinted at in Model 3.

Uses related to decay

Many techniques for dating materials rely on the decay of radioactive materials. The classic example is radiocarbon dating. A tiny fraction of the carbon in the environment is a radioactive form, carbon-14. This decays to form nitrogen-14. Living creatures contain carbon-14 in the same proportion as in the environment. However, when they die, no new carbon-14 enters their bodies, and that which is present gradually decays. So, in principle, the older a sample of once-living material, the smaller the fraction of carbon-14 that is present.

Care must be taken to allow for variations in the level of carbon-14 historically, and for possible contamination. The burning of coal in large quantities has introduced much 'old' carbon into the environment, and this has affected the proportion of carbon-14 in objects which have died since the industrial revolution.

The half-life of carbon-14 is about 5570 years, so its use is limited to objects between 500 and 25 000 years old, approximately. Other radioisotopes can be used for older objects. In particular, rocks can be dated by measuring the ratio of particular isotopes of potassium and argon.

Further activities

♦ Pupils could research the Turin Shroud. Its dating (using carbon-14) to the Middle Ages is still hotly debated. A detailed discussion of the evidence relating to the shroud by one of the scientists who developed the sensitive dating techniques used on it is given in *Relic, Icon or Hoax?* (Gove, 1996).

♦ School radioactive sources may well have been bought a decade or more ago. This means that they will be less active than when bought. In particular, cobalt-60 (a gamma source) has a half-life of just over five years. If you can find out when it was bought, pupils could estimate what fraction of its activity remains. (Every school should have a record showing when each source was obtained.)

Enhancement ideas

♦ The half-lives of radioactive elements vary widely, from fractions of microseconds to billions of years.

♦ Radioactive elements with short half-lives are used in medicine. For example, in a gamma scan, the patient is injected with a gamma-emitting substance. Because it decays rapidly, it gives out its radiation quickly and so only a small amount is required.

Short half-life radioisotopes for medical and other uses are prepared by exposing non-radioactive elements to high-energy neutrons or protons.

♦ Radiocarbon dating of the Iceman, whose body was discovered in an Alpine glacier, showed that he had lived thousands of years earlier than predicted from his appearance and possessions.

6.3 Explaining radioactivity

In chemistry, pupils experiment with many different substances ('chemicals'). They have to learn that, at the end of a chemical reaction, the same elements and indeed the same atoms are present as were there at the start. Although the substances may change colour, state and so on, all that has happened is a rearrangement of atoms. Radioactivity is different. During radioactive decay, the very atoms change.

We all know that medieval alchemists tried to turn lead into gold. Whilst this is a gross oversimplification, it can be helpful to pupils to point out that what the alchemists were attempting was impossible; they were trying to use chemical reactions to transmute elements, when what is needed is a nuclear reaction.

◆ *Previous knowledge and experience*

The extent that you will have to deal with the atomic nature of matter and the structure of atoms will depend on what pupils have studied in chemistry. Careful co-ordination is needed here.

◆ *A teaching sequence*

We will assume that pupils are familiar with the idea that matter is made of atoms and molecules. You may wish to reinforce this by looking at some recent images which show matter magnified a billion times – images from scanning tunnelling electron microscopes or atomic force microscopes. These can show individual atoms packed together to make solids; they can also show large molecules (such as DNA) as blobs or 'snakes'. In the past, science teachers used to tell their classes that we could only imagine atoms, that they were too small ever to be seen. These modern techniques have given the lie to this, and it is worth making full use of the images they produce. More able pupils might find out how these techniques work. You could also discuss whether they really allow us to 'see' atoms.

Atoms and nuclei

The first evidence for the existence of the atomic nucleus came from what is known as Rutherford's alpha particle scattering experiment. The experiments were actually performed by Geiger and Marsden. They directed a beam of alpha particles at thin gold foil. To their surprise, they found that a small fraction were scattered back – they had 'bounced' off something in the sample.

This was surprising because at that time atoms were pictured as spheres of positive charge with electrons embedded in them (the 'plum-pudding' model). Positively charged alpha particles should have passed through almost unaffected. Rutherford proposed that all the positive charge must be concentrated in a tiny nucleus, and he was able to estimate its size. This model seemed unlikely at the time, because positive charges repel one another, so the nucleus should be intrinsically unstable. (Later, an additional nuclear force was identified which holds the nucleus together against the electrostatic repulsion.)

You could describe this experiment, or show a suitable video (see *Other resources*, page 253). Pupils could discuss the reasons for abandoning the plum-pudding model.

They could also investigate some 'mystery boxes' (or make their own). These are boxes with greaseproof paper sides, containing mystery objects. They can only be investigated by shining light on them; from the shadows, can the objects be identified? In an alternative version, the boxes have black paper sides and you are only allowed to poke knitting needles through them.

You might demonstrate an 'alpha scattering analogue' (Figure 6.7) – an aluminium or plastic hill towards which you roll marbles. The marbles follow tracks similar to those of alpha particles near a gold nucleus. A few score direct hits and are scattered back towards where they came from.

Figure 6.7
An analogue of alpha scattering.

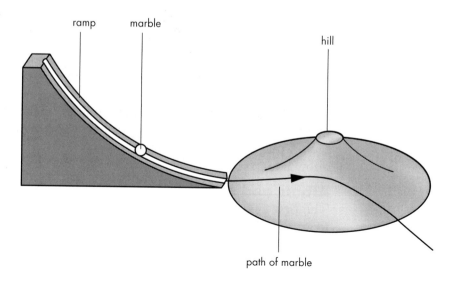

ramp   marble   hill

path of marble

In another analogue, a metallised polystyrene ball hangs on a nylon thread, level with the top dome of a Van de Graaff generator. Touch the ball on the charged Van de Graaff dome to charge it up. Now pull it back and let it swing towards the dome. It will be repelled (because like charges repel). Its path is a good model for the path of an alpha particle near a gold nucleus.

Neutrons, protons and electrons

We will assume that pupils are familiar with the idea that atoms are made of protons, neutrons and electrons. They should also be aware from their study of chemistry that an element is characterised by the number of protons in the nucleus of each atom; atoms of an element with different numbers of neutrons are isotopes of that element. (The term *radioisotope* simply refers to an isotope of any element that is radioactive.)

This is a suitable point to introduce the underlying natures of alpha, beta and gamma radiations. If pupils are familiar with protons, neutrons and electrons, they can extend this to the three types of radiation as shown in Table 6.1. Given the first three columns, they should be able to work out the charge and mass of each type.

Table 6.1

Particle	Symbol	Same as ...	Charge	Mass
proton	p		+1	1
neutron	n		0	1
electron	e		−1	0
alpha	α	2p and 2n	+2	4
beta	β	e	−1	0
gamma	γ	electromagnetic radiation	0	0

You may need to emphasise that these values of charge and mass are relative, and that the relative masses are approximate.

You may also need to emphasise that all three types of radiation are emitted *by the atomic nucleus*. Beta radiation takes the form of an electron emitted from the nucleus; it is *not* one of the orbital electrons. In effect, a neutron decays to give a proton and an electron:

$$n \longrightarrow p + e$$

Pupils could check that both mass and charge are conserved in this reaction. Crudely speaking, we can say that beta particles are emitted by nuclei that have 'too many neutrons', while alpha particles are emitted by those with 'too many protons'. Gamma rays are emitted by nuclei that are left with an excess of energy after emitting an alpha or beta particle.

Ionisation

Radiation causes ionisation by knocking one or more electrons from an atom as it passes. A single alpha, beta or gamma ray may ionise many atoms – perhaps hundreds of thousands – before it loses all its energy. Now that you have described how radiation is emitted by atomic nuclei, you can describe how it can ionise atoms as it whizzes past (Figure 6.8).

Figure 6.8
*Ionisation of atoms by a passing alpha particle.
(Adapted from: Basic Physics 1 and 2 by UCLES, 1995. Cambridge University Press.)*

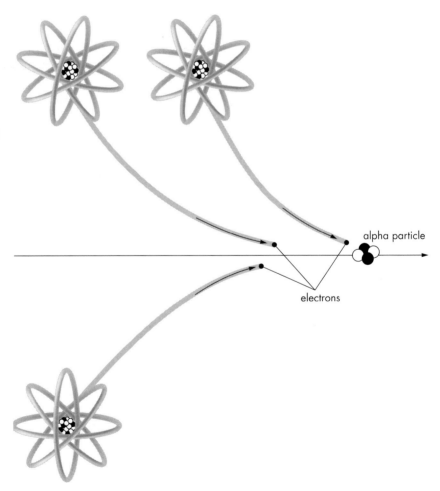

alpha particle

electrons

Alpha radiation has the greatest ionising effect; it can be likened to a lumbering cannon ball, compared to the high-velocity bullet which is a beta particle. Consequently alpha radiation is the most damaging but the least penetrating (because its energy is used up in the shortest distance). At this point, you could recall some of the phenomenological points discussed in Section 6.1, and show how they can be explained in terms of the mass, charge and velocity of the radiation.

◆ *Further activities*

If your syllabus requires you to consider the discovery of the electron, you will find some relevant activities in *One Hundred Years of the Electron* (Sang, 1997).

◆ *Enhancement ideas*

◆ It is an oversimplification to suggest that there are only three types of nuclear radiation, α, β and γ. Some nuclei emit positive beta radiation (a form of anti-matter), whilst others emit neutrons.

◆ During beta decay, a further particle is emitted – the neutrino. This odd little particle has almost no mass; it has no charge but (like an electron) it does have spin. Some cosmologists believe that neutrinos may account for 90% of the mass of the Universe.

◆ As far as anyone can tell, both electrons and neutrinos are fundamental particles; they are not composed of any other sub-particles, and they appear to have no size!

6.4 Fission, fusion and energy

The development of nuclear weapons and that of nuclear power stations have been closely linked. Much of the science and technology are the same, and some of the first civil nuclear power stations were designed to produce plutonium for use in weapons. The international community is cautious about the spread of nuclear power station technology because of the link with nuclear weapons. Pupils are conscious of this link, although it is not generally acknowledged in syllabuses. You may wish to acknowledge it in your teaching.

♦ *Previous knowledge and experience*

Pupils have often learnt about nuclear power stations as 'another type of power station', and they are likely to be aware that they use uranium as a fuel and that they produce dangerous, long-lived waste.

♦ *A teaching sequence*

The mechanism for controlling nuclear fission in a reactor is difficult to grasp, so you may prefer to start with a consideration of fission bombs.

Nuclear fission

Fission is a form of stimulated radioactive decay. A neutron strikes a large nucleus (such as uranium or plutonium) causing it to split. Energy is released, along with two or three more neutrons. If at least one of these neutrons strikes a further nucleus, a chain reaction develops. Several videos are available which show an animated version of this process, usually in connection with nuclear power.

To ensure that a chain reaction starts, a minimum mass of uranium is needed. With less than this 'critical mass', neutrons simply escape without causing further fission. In a fission bomb, two sub-critical masses are brought together to start a chain reaction. A source of neutrons may also be needed to trigger the reaction since spontaneous fission is rare.

There should be no mystery about the energy released in fission: a small fraction (less than 10%) is in the form of gamma radiation; the rest is in the form of kinetic energy of the particles which result. Pupils should picture the various particles (daughter nuclei, neutrons, beta particles) recoiling in different directions from the site of the parent nucleus. They collide with surrounding uranium atoms and share their energy. So the atoms of the uranium gain kinetic energy; in other words, the uranium gets hot. Thus it is best to avoid talking about 'nuclear energy' as if it were some different form of energy.

It is important to emphasise that, although we talk of uranium as a fuel, and even about uranium being 'burned', the process is very different from oxidation. Nuclear bonds are breaking and reforming, rather than interatomic bonds. The energy released in each fission event is perhaps a million times greater than that released during the oxidation of a hydrogen molecule.

Modelling chain reactions

Pupils will be familiar with the idea of a chain letter. They will appreciate the fact that not everyone passes the letter on, so that the postal service is not overburdened by an exponential growth in letters; in the same way only some of the neutrons released in one fission event go on to stimulate others.

Make a line of equally spaced dominoes, standing on end. Tip the first one over, and watch the line topple. This represents a controlled chain reaction. Now introduce some 'forks' into the line, to represent a chain reaction that runs out of control.

Another model which is rather dramatic uses matches. Low down on a vertical board, stick a single unused match. With this match as the lower vertex, make a triangle of matches (Figure 6.9). Light the first one; this ignites the two above, which ignite the three above that, and so on.

 Carry this out only as a demonstration and in a safe area, perhaps outside.

Figure 6.9
Modelling a chain reaction.

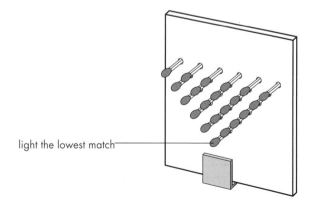

light the lowest match

 Pupils could devise computer models to represent chain reactions.

Nuclear power stations

Again, a video is likely to be the best way to approach this topic. While watching the video (or afterwards), pupils could be asked to label a diagram of a nuclear power station. At a simple level, pupils should appreciate that uranium gets hot as a result of fission, and this heat is used to boil water as in a fossil-fuel power station. A more sophisticated understanding requires pupils to appreciate the role of control rods in maintaining the desired level of neutrons, and of the moderator in slowing down neutrons to ensure that they interact with the uranium nuclei.

Nuclear waste

Another way for pupils to compare fossil-fuel and nuclear power stations is as black boxes. They should identify inputs and outputs; this will highlight the similarities, and also open up the question of nuclear waste. This gives pupils an opportunity to use their understanding of radioactivity and its effects on living organisms – they will have to use the ideas of penetrating power, cell damage and half-life.

Assuming that no-one wants a nuclear waste dump in their back yard, pupils could discuss the 'NIMBY' argument (Not In My Back Yard). To what extent is it reasonable to expect one part of society to live in close proximity to everyone's nuclear waste?

Nuclear fusion

Fusion is the process whereby atoms of all elements have been built up from protons, neutrons and electrons. In the immediate aftermath of the Big Bang (see Chapter 5), only the three lightest elements were produced – hydrogen, helium and lithium. Subsequently, heavier elements were formed in the hot cores of stars. The heaviest elements were formed in giant stars which ended their lives as supernovas (stellar explosions), casting vast amounts of material into space. It is from this material that the Solar System formed, so we can be said to have formed from stardust. The energy released by radioactive materials today is the remnant of the energy which those materials acquired during the last stages of a star's life.

Pupils could make a chart or cartoon to show how the material of which their bodies is made originated in the Big Bang. Suitable stimulus material can be found in *World of Science* (ASE, 1997). A history of the very stuff of which we are made is perhaps a suitable point at which to conclude a study of radioactivity.

◆ *Further activities*

- ◆ Radioactive decay within the Earth releases energy which keeps the inside of the Earth fluid and powers continental drift. An activity relating to this can be found in *Henri Becquerel and the Discovery of Radioactivity* (Sang, 1997).

- ◆ The nuclear industry produces a wealth of information leaflets which are freely available to schools. These can be used in different ways: for example, as a research resource for pupils, or as material for critical analysis – how objective do pupils consider them to be?

- ◆ There are many reactors which are not used for generating electricity. For example, some are used for producing radioisotopes for medical or other uses. Pupils could research these uses.

◆ *Enhancement ideas*

- ◆ There is evidence that, about two billion years ago, a rich seam of uranium-bearing rock in Gabon (West Africa) became a natural fission reactor.

- ◆ The first controlled nuclear chain reaction was set up by Enrico Fermi in a disused squash court at the University of Chicago in 1942. He, at least, was confident that the reaction would remain under control.

- ◆ The uranium fuel rods used in nuclear power stations represent little hazard. They are transported in conventional trucks, and the rods themselves can be handled safely by workers wearing protective gloves. The spent fuel rods are more hazardous, and the concentrated waste extracted from them is extremely so.

⚠ 6.5 Handling radioactive sources

This section looks at the radioactive sources available for use in schools, and considers how they may be handled safely. Although it is intended to help you, the teacher, it may also help you to answer some of your pupils' questions.

Low-level sources

You do not need to have special authorisation to buy or handle specimens of rock which exhibit low but measurable levels of radioactivity. Indeed, you may be able to find some of your own in suitable parts of the country.

You may also have an old luminous clock which has radioactive paint on its hands. You can also buy mantles for camping gas lamps; these are radioactive because they contain thorium.

It is useful to have a collection of such items to emphasise the point that radioactive materials and their radiation are all around us, and that not all radioactive sources are so dangerous that we have to take strict precautions.

Sealed sources

It is worth understanding how typical sealed sources are constructed (Figure 6.10). The radioactive material is added to metal and made into a foil. This is then mounted in a metal holder with gauze across the open end. The gauze prevents you from putting your fingers inside, but allows radiation out. Don't forget that the penetrating power of gamma radiation means that it can emerge through the metal holder.

Figure 6.10

A typical laboratory radioactive source.

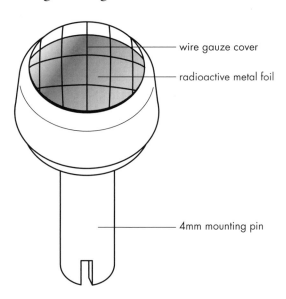

wire gauze cover

radioactive metal foil

4mm mounting pin

Table 6.2 *Sources available for use in schools.*

Source	Radiation emitted	Half-life
americium-241	alpha and gamma	433 years
strontium-90	beta	29 years
cobalt-60	gamma	5.27 years
radium-226	alpha, beta and gamma	1620 years

Table 6.2 shows the sources that are most likely to be available to you. It is simplest to refer to them in class as 'the gamma source' etc., rather than 'the cobalt-60 source' etc., since these names will be unfamiliar to pupils. The notional activity of each of these sources when supplied is 185 kBq (or 5 microcuries, μCi), but remember that this will have decreased since the source was bought.

These sources are supplied in lead-lined boxes for safe keeping, but remember that some radiation can escape through the box. The sources should be handled with tongs and wherever possible they should be taken from their box and inserted directly into a holder standing on the bench. The stem of the source is 4 mm in diameter and suitable holders can be bought; they can also be made, since the stem fits neatly into a 4 mm electrical socket. Using a holder ensures that you do not have to spend much time with the source in close proximity to your body, or to your pupils.

It is a good idea to place a source in a mount and then to use a Geiger counter to explore the strength of the radiation around it. You should find that, further than about a metre from the source, the level of the radiation is close to the background level, and so you and your pupils will be safe if you spend most of your time at least a metre from the source.

Other sources

The protactinium generator has been discussed above, in Section 6.2. Other sources require special authorisation from the DfEE.

Recording use

You are legally obliged to keep a record of the times of your exposure to radioactive sources and of any of your pupils aged over 16; younger pupils may only observe these experiments. You need to record four things, from which radiation doses could, in principle, be calculated:

- the activity of the source;
- the type of radiation emitted (α, β, γ);
- the time for which you were exposed;
- how close you were to the source.

If you work carefully, as described above, the closest you will come to a source will be about 10 cm (when you are handling sources with tongs) and you can probably limit the time of your exposure to 5 minutes or so. (Most of the time the sources will be mounted in holders and you will be at a safe distance.) The dose of radiation you then receive will be trivial compared with your exposure to background radiation over a year.

Bear in mind the self-explanatory mantra for safety in the nuclear industry: *time, distance, shielding.*

A last word of advice

There is a minimal hazard associated with working with the radioactive sources described above, provided you follow the practices suggested. Try not to suggest to your pupils that you are doing anything very dangerous. When you are performing a demonstration, emphasise the way in which you are working to minimise the risk to yourself and to them. This will help them to understand the physics of radioactivity, as well as encouraging them to take a rational view of its hazards.

◆ *References*

ASE, 1991: *SATIS* Unit **1105b** *Radon – an investigation*. ASE publications.

ASE, 1997: *World of Science*. John Murray/ASE.

Gove, Harry E., 1996: *Relic, Icon or Hoax?* Institute of Physics Publishing.

Sang, David (ed.), 1997: *Henri Becquerel and the Discovery of Radioactivity*. ASE publications.

Sang, David (ed.), 1997: *One Hundred Years of the Electron*. ASE publications.

◆ *Other resources*

- ◆ The National Radiological Protection Board (Chilton, Didcot, Oxon, OX11 0RQ) produces many public information leaflets relating to radiation and its effects.
- ◆ *SATIS* (Science and Technology in Society) materials for 14–16 year olds are published by the ASE. Relevant units include: **204** Using radioactivity, **1105** Radon in homes.
- ◆ Becquerel's experiments are discussed in *Exploring the Nature of Science*, J. Solomon, ASE.
- ◆ Half-lives of many radioisotopes are listed in *Revised Nuffield Advanced Science: Book of Data*, ed. H. Ellis, Addison Wesley Longman.
- ◆ A video, *Discovery and the Atom*, describes Rutherford's work (with archive film) and discusses the nature of scientific models such as the nuclear model of the atom. Originally produced by UKAEA.
- ◆ You could contact environmental organisations, such as Greenpeace or Friends of the Earth, for samples of their publications relating to nuclear issues.

Visits
The Science Museum in London has a gallery devoted to nuclear power.

Background reading
For a discussion of a general approach to this topic and how it can be related to more general ideas about radiation, see:
Millar, R., Klaassen, K. and Ejkelhof, H.: 'Teaching about radioactivity and ionizing radiation: an alternative approach', in *Physics Education* 25 338, 1990.

Extracts from Becquerel's diaries can be found in:
Carey, J. (ed.), 1995: *The Faber Book of Science*. Faber and Faber.

For a more advanced treatment of the physics of fission and fusion and how these processes are used in weapons and power stations, see:
Sang, David, 1995: *Nuclear and Particle Physics*. Thomas Nelson.

The work of Marie and Pierre Curie, and the development of radiochemistry, are described in:
Ellis, P. (ed.), 1999: *100 years of Radium*. ASE.

Index

Page references in *italics* refer to Figures; those in **bold** refer to Tables

INDEX